CHILTON'S REPAIR & TUNE-UP GUIDE
PLYMOUTH
1968 to 1976

**Fury 1,11,111 • Sport Fury • VIP • Suburban
Gran Coupe • Gran Sedan • Gran Fury**

Managing Editor KERRY A. FREEMAN, S.A.E.
Senior Editor RICHARD J. RIVELE, S.A.E.

President WILLIAM A. BARBOUR
Executive Vice President JAMES A. MIADES
Vice President and General Manager JOHN P. KUSHNERICK

CHILTON BOOK COMPANY
Radnor, Pennsylvania
19089

SAFETY NOTICE

Proper service and repair procedures are vital to the safe, reliable operation of all motor vehicles, as well as the personal safety of those performing repairs. This book outlines procedures for servicing and repairing vehicles using safe, effective methods. The procedures contain many NOTES, CAUTIONS and WARNINGS which should be followed along with standard safety procedures to eliminate the possibility of personal injury or improper service which could damage the vehicle or compromise its safety.

It is important to note that repair procedures and techniques, tools and parts for servicing motor vehicles, as well as the skill and experience of the individual performing the work vary widely. It is not possible to anticipate all of the conceivable ways or conditions under which vehicles may be serviced, or to provide cautions as to all of the possible hazards that may result. Standard and accepted safety precautions and equipment should be used when handling toxic or flammable fluids, and safety goggles or other protection should be used during cutting, grinding, chiseling, prying, or any other process that can cause material removal or projectiles.

Some procedures require the use of tools specially designed for a specific purpose. Before substituting another tool or procedure, you must be completely satisfied that neither your personal safety, nor the performance of the vehicle will be endangered.

Although the information in this guide is based on industry sources and is as complete as possible at the time of publication, the possibility exists that the manufacturer made later changes which could not be included here. While striving for total accuracy, Chilton Book Company cannot assume responsibility for any errors, changes, or omissions that may occur in the compilation of this data.

PART NUMBERS

Part numbers listed in this reference are not recommendations by Chilton for any product by brand name. They are references that can be used with interchange manuals and aftermarket supplier catalogs to locate each brand supplier's discrete part number.

ACKNOWLEDGMENTS

The Chilton Book Company expresses its appreciation to the Chrysler-Plymouth Division, General Motors Corporation, Detroit, Michigan and Lee Templeton Motors, Norristown, Pennsylvania for their generous assistance.

Manufactured in the United States of America
567890   543210

Chilton's Repair & Tune-Up Guide: Plymouth 1968–76
ISBN 0-8019-6551-9
ISBN 0-8019-6552-7 pbk.
Library of Congress Catalog Card No. 76-9517

CONTENTS

Quick Reference
Specifications For Your Vehicle

Fill in this chart with the most commonly used specifications for your vehicle. Specifications can be found in Chapters 1 through 3 or on the tune-up decal under the hood of the vehicle.

Tune-Up

Firing Order_____

Spark Plugs:

    Type_____

    Gap (in.)_____

Point Gap (in.)_____

Dwell Angle (°)_____

Ignition Timing (°)_____

    Vacuum (Connected/Disconnected)_____

Valve Clearance (in.)

    Intake_____   Exhaust_____

Capacities

Engine Oil (qts)

    With Filter Change_____

    Without Filter Change_____

Cooling System (qts)_____

Manual Transmission (pts)_____

    Type_____

Automatic Transmission (pts)_____

    Type_____

Front Differential (pts)_____

    Type_____

Rear Differential (pts)_____

    Type_____

Transfer Case (pts)_____

    Type_____

FREQUENTLY REPLACED PARTS
Use these spaces to record the part numbers of frequently replaced parts.

PCV VALVE	OIL FILTER	AIR FILTER
Manufacturer_____	Manufacturer_____	Manufacturer_____
Part No._____	Part No._____	Part No._____

General Information and Maintenance

HOW TO USE THIS BOOK

Chilton's Repair and Tune-Up Guide for the Plymouth is designed for the owners of full-sized Plymouths who want to do some of the service on their own cars. Included are step-by-step instructions for maintenance, troubleshooting, repair or replacement of many components of the car.

Model years from 1968 to 1976 are covered by this guide and the engines included are the 225 cu in. slant six, and the following V8s: 318, 360, 383, 400, and 440 cu in. displacement engines.

Before attempting to perform any of the service procedures outlined in this repair and tune-up guide, thoroughly familiarize yourself with the steps of the procedures you are going to perform. Read each step carefully, being sure to note the tools that you will need for each procedure.

Safety is an important factor when performing any service operation. Areas of special hazard are noted in the text, but common sense should always prevail.

Here are some general safety rules:

1. When working around gasoline or its vapors, don't smoke and remember to be careful about sparks which could ignite it.

2. Always support the car securely with jackstands (not milk crates!) if it is necessary to raise it. Don't rely on a tire changing jack as the only means of support. Be sure that the jackstands have a rated load capacity adequate for your car.

3. Block the wheels of the car which remain on the ground, if only one end is being raised. If the front end is being raised, set the parking brake as well.

4. If a car equipped with an automatic transmission must be operated with the engine running and the transmission in gear, always set the parking brake and block the *front* wheels.

5. If the engine is running, watch out for the cooling fan blades. Be sure that clothing, hair, tools, etc., can't get caught in them.

6. If you are using metal tools around or if you are working near the battery terminals, it is a good idea to disconnect it.

7. If you want to crank the engine, but don't want it to start, *remove* the high tension lead which runs from the coil to the distributor.

TOOLS AND EQUIPMENT

Before attempting service procedures, it is important to have at least some basic tools and equipment on hand. Tools are easy to obtain for Plymouths as they use standard SAE nuts, bolts, and fittings.

Remember to use the proper tool for each nut, bolt, or fitting. If a torque wrench is called for and torque specifications given, use them.

Be sure not to force things. If a part won't come off, a hammer and a pair of locking pliers is not the solution; patience and care are. It is much easier to do something right the first time than to repeat it or replace a damaged part.

The following is a recommended list of basic tools and equipment for anyone wishing to perform maintenance or tune-ups on their car:

Tools:

- Assorted screwdrivers
- Assorted allen keys
- Socket wrench set (include spark plug socket)
- Torque wrench
- Assorted crescent wrenches (or several adjustable wrenches)
- Oil filter (band) wrench
- Wire feeler gauges (spark plugs)
- Flat feeler gauges (points)
- Pliers

Equipment:

- Hand-operated grease gun
- Timing light
- Dwell/tachometer
- Tire pressure gauge
- Vacuum gauge
- Remote starting switch
- Jackstands and wheel blocks

If you wish to perform more complex service procedures, you will probably find it necessary to buy or borrow more tools. Check the procedure before beginning it, to see what tools are necessary.

MODEL IDENTIFICATION

1968 Fury

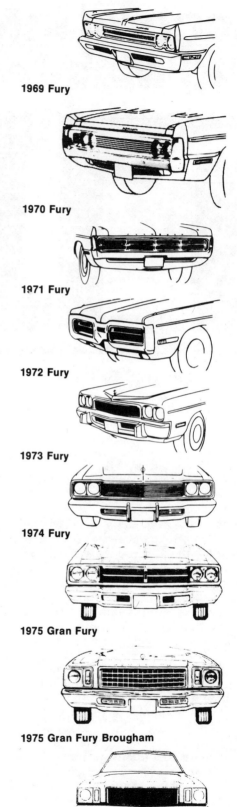

1969 Fury

1970 Fury

1971 Fury

1972 Fury

1973 Fury

1974 Fury

1975 Gran Fury

1975 Gran Fury Brougham

1976 Gran Fury

Vehicle Identification

First Digit— Series	Second Digit— Price Class	Third and Fourth Digits—Body Style	Fifth Digit— Engine (cu in.)	Sixth Digit— Model Year	Seventh Digit— Assembly Plant
P—Plymouth Fury (full size)	E Economy L Low M Medium H High P Premium T Taxi S Special O Superstock	21 2-dr. sedan 23 2-dr. hardtop 27 Convertible 29 2-dr. special 41 2-dr. sedan 43 4-dr. hardtop 45 2-seat wagon 46 3-seat wagon	See Engine Code Chart	8 1968 9 1969 0 1970 1 1971 2 1972 3 1973 4 1974 5 1975 6 1976	A Lynch Rd. B Hamtramck C Jefferson D Belvidere E Los Angeles F Newark G St. Louis P Wyoming (export) R Windsor

SERIAL NUMBER IDENTIFICATION

Vehicle

The vehicle identification number (VIN) is stamped on a metal plate which is attached to the left-side of the instrument panel. The plate is visible through the windshield.

The VIN contains 13 digits. The first digit shows the series; the second, the price class; the third and fourth, the body style; the fifth digit, engine size; the sixth digit, year; and the seventh digit, assembly plant. The remaining six digits are the car's sequential serial number.

The VIN plate is located on the top of the instrument panel

Engine

The engine code is the fifth number of the vehicle identification number (VIN).

Engine Code

Disp	Bbl	Hp	'68	'69	'70	'71	'72	'73	'74	'75	'76
6-Cylinder Models											
225	1	145	B	B	C	C					
8-Cylinder Models											
318	2	150, 170 (net) °					G	G		G	G
318	2	230	F	F	G	G					
360	4	240 (net)							J	J	
360	2	160,170, 180 °					K	K	K	K	K
383	2	270									
383	2	275				L					
383	2	290	G	G	L						
383	4	300				N					
383	4	325									
383	4	330	H	H	L						
383	4	335	H	H	N						
400	2	190 (net)					M				
400	2	175 (net)						M	M	M	M
400LB	4	210									N
400	4	255 (net) ①					P				
400HP	4	260 (net)						P	P		
440	4	205, 215, 220, 225 °					T	T	T	T	T
440HP	4	275, 280 (net) °						U	U		
440HP	4	290 (net)					U				
440	4	370				U					
440	4	375	L	L	U						
440	6	385				V					
440	6	390			V						

LB Lean Burn
° Net horsepower rating varies with model application
① Non-California cars with Fresh Air Pack—265 (net)

The VIN is on the left-hand side of the instrument panel where it is visible through the windshield. The VIN is also located to the rear of the right engine mount on 1969–76 models.

VIN code location—6 cyl

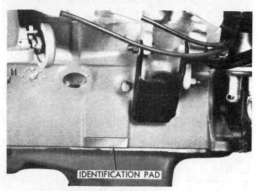

VIN code location—V8

Transmission Serial Numbers

MANUAL TRANSMISSION

The serial numbers for all manual transmissions are stamped on a pad on the right-side of the transmission case. Using PP 833 1861 0275 as an example, the serial number is interpreted as follows: manufacturing plant (first and second letters), transmission model number (third, fourth, and fifth digits), manufacturing date code (sixth through the ninth digits), and production sequence number (tenth through the thirteenth digits).

Manual Transmission

Type	Model Number	Application
3-speed	A-903	6 cylinder models 1968–71
3-speed	A-745	8 cylinder models and 6 cylinder heavy-duty use 1968–69
3-speed	A-230	8 cylinder models and 6 cylinder heavy-duty use 1970–71
4-speed	A-833	8 cylinder models 1968

AUTOMATIC TRANSMISSIONS

The automatic transmission identification code and serial number are cast in raised letters on the lower left-side of the bellhousing.

Automatic Transmissions

TorqueFlite	Usage
A-904-G	6 cylinder engines 1968–71
A-904-LA	V8-318 engines①
A-727-A	V8-318 and 360 engines①
A-727-B	V8-383, 400, and 440 engines

① V8-318 cu in. engines use either A-727-A or A-904-LA transmission interchangeably
NOTE: *Transmission version used is dependent upon engine, coupled with probable usage.*

Manual transmission ID

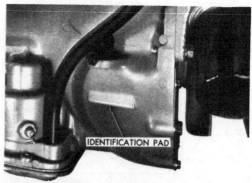

Automatic transmission ID

ROUTINE MAINTENANCE

Air Cleaner

The air cleaner element used on the Plymouth is of the disposable, dry-paper type. It should be checked every 6 mos/6,000 miles, and replaced every 2 yrs/24,000 miles (30,000 miles—1975–76), whichever comes first.

However, if the vehicle is used in a very dusty area, or if it has a Fresh Air Induction System, the air cleaner should be inspected at every oil change and replaced once a year.

To remove and inspect the air cleaner, proceed in the following manner:

1. Disconnect the PCV hose at the air cleaner end. Undo the wing nut and lift the air cleaner from the carburetor. Remove the paper element from the air cleaner.

2. Check the element. If it is extremely dirty, or if it is saturated with oil for more than one-half of its circumference, replace it. In the case of oil saturation, check the rest of the PCV system to be sure that it is working properly.

Cleaning the air cleaner element

3. Clean the element with compressed air by holding the nozzle 2 in. or more from the inside of the screen.

NOTE: *Cleaning the element from the outside of the screen can result in foreign material becoming embedded in it.*

4. Check the element for improper sealing, punctures, etc. Installation is the reverse of removal.

PCV System

The positive crankcase ventilation (PCV) valve and system should be checked every 6 months (1968–71) or every 12 months (1972–76). The valve should be replaced at the following intervals:

- 1968–71—every year
- 1972–74—every two years
- 1975–76—every 30,000 miles

For PCV valve testing, as well as removal, see Chapter 4.

At the same time the PCV system is being checked, the crankcase inlet air cleaner should be cleaned:

1. Remove the hose which runs to the carburetor air cleaner from the crankcase inlet air cleaner (breather cap).

The crankcase (PCV) air cleaner assembly

2. Inspect the hose and clean or replace it, as necessary.

3. Remove the crankcase air cleaner assembly from the valve cover.

4. Wash the assembly in kerosene.

5. Invert the assembly and wet the filter element by filling it with SAE 30 engine oil.

6. Allow the excess oil to drain off and install it on the engine. Connect the hose back to the air cleaner.

Charcoal Canister

1972–76 Models

Inspect the charcoal canister and replace the element in its base at the following intervals:

1972–74—every 12 mos/12,000 mi
1975–76—every 15,000 mi

If the car is driven under dry, dusty conditions, replace the filter more often.

To replace the filter element, proceed in the following manner:

1. Loosen the screws which secure the canister bracket and lift out the canister.

2. Invert the canister and lift out the old filter element.

Changing the filter in the charcoal canister

3. Work the new filter into place with your fingers.

4. Install the canister and tighten its retaining screws.

5. Check the vacuum hoses to be sure that they are not loose, pinched, or clogged and make sure that they are routed as marked on the canister top.

Accessory Drive Belts

Drive belts should be checked routinely; for example, at every oil change. Check all drive belts for cuts, cracks, or other signs of wear, and replace them as required. Be sure that there is enough clearance between the belts and other engine components. Check belt tension and adjust it as outlined below.

TENSION CHECKING AND ADJUSTING

1. Push down on the accessory belt, which is being checked, at the midpoint of its longest section.

2. With moderate thumb pressure, the belt should deflect ⅜–½ in.

3. If necessary, loosen the adjusting link bolt and move the accessory until proper belt tension is obtained. For some accessories a ½ in. square hole is provided on the bracket for this purpose;

use a ratchet handle with a ½ in. drive to apply pressure to the accessory.

CAUTION: *Do not pry on accessories with aluminum housings.*

4. Tighten the adjusting link bolt and recheck belt deflection.

Air Conditioning

SIGHT GLASS CHECKING

1968–73

The ambient temperature must be between 70°–110°F in order to perform this test.

1. Start the car and allow it to reach normal engine idle speed and operating temperature.

2. Turn the air conditioner on and set the blower switch on the "High" position. If the car has dual system, operate both blowers at high speed.

3. Open all of the car's doors and windows.

4. Look at the sight glass, which is located on the top of the receiver/drier in the engine compartment.

5. There should be no foam or bubbles visible in the sight glass. If there are, have the system checked.

6. If the air conditioner is not cooling at all, and the sight glass is clear, there is probably no refrigerant in the system.

7. Check the air conditioning lines and fittings for signs of leakage; oil visible on them indicates a leak. Do not attempt to tighten fittings or replace lines yourself.

CAUTION: *Your automobile air conditioner has no owner serviceable parts. Contact with refrigerant could*

The sight glass is located on the top of the receiver/drier

cause severe injury or burns. Improper service could result in an explosion. For these reasons, air conditioner service must be left to qualified service personnel.

1974–76

The ambient temperature should be between 70°–110°F for this test.

1. Start the engine; place the controls on "A/C", the fan switch on "High", and the temperature selector on "Cool".

2. Run the engine for several minutes.

3. Look at the sight glass, which is located on the receiver/drier in the engine compartment. It should be clear. If it is clear, the inlet line to the compressor cold, and the outlet line warm, everything is in good working order.

4. If the sight glass is clear, the compressor clutch engaged, and both compressor lines the same temperature (warm), the refrigerant is completely discharged from the system.

5. If the sight glass is clear and the compressor clutch is disengaged, the compressor or its wiring is faulty.

6. If there are foam or bubbles visible in the sight glass, the refrigerant is probably low. Increase the engine speed to 1,600 rpm on sixes, or 1,300 rpm on V8s, if the bubbles remain, the refrigerant is low.

NOTE: *If the ambient temperature is below 70° or above 110° F, some bubbling or foaming is normal.*

7. Check the A/C system lines and fittings for traces of oil. If any oil is visible, the system has a leak. Do not attempt to replace lines or tighten fittings yourself.

8. If any of the above checks indicate a defect, have the system checked by qualified air conditioning service personnel.

CAUTION: *Your automobile air conditioner has no owner serviceable parts. Contact with refrigerant can cause severe injury or burns. Improper service could result in an explosion. For these reasons air conditioner service must be left to qualified service personnel.*

Periodic Operation

During colder months when the air conditioner is not in use, Plymouth rec-ommends that you operate the system once a week for five minutes.

Set the temperature lever on "Warm" and depress the "MAX-A/C" button.

This will allow the compressor to pump lubricant through the system to prevent the seals from drying.

Fluid Level Checks

ENGINE OIL

The engine oil level should be checked at regular intervals; for example, whenever the car is refueled. Check the oil level if the red oil warning light comes on.

It is preferable to check the oil level when the engine is cold or after the car has been standing for a while. Checking the oil immediately after the engine has been running will result in a false reading. Be sure that the car is on level surface before checking the oil level.

Remove the dipstick and wipe it with a clean rag. Insert it again (fully) and withdraw it. The oil level should be between the "FULL" and the "ADD OIL" marks. Do not run the engine if level is below the "ADD OIL."

Add oil, as necessary. Use oil which carries the API designation SE.

CAUTION: *Do not use unlabeled oil or a lower grade of oil which does not meet SE specifications.*

Do not overfill; the oil level should never be above the "FULL" mark.

TRANSMISSION

Manual

The fluid level should be checked every six months and should also be replenished, if required. The oil level should reach the bottom of the filler plug. When additional oil is necessary, add DEXRON® automatic transmission fluid. In warm climates, Multi-purpose Gear Lubricant, SAE 90 may be substituted in the three-speed transmission.

If gear rattle at idle becomes annoying, in the four-speed units, drain the factory fluid and replace it with Multipurpose Gear Lubricant, SAE 140.

Automatic

The transmission fluid level should be checked every six months. Check it after

Manual transmission drain and fill plug location

the car has reached normal operating temperatures. Apply the handbrake and allow the engine to idle. Briefly engage each gear, finishing in "N" (Neutral). The level on the transmission dipstick should be at the "F" mark, or just below it; it should never register above it.

Add fluid, if it is required, through the dipstick hole. The correct fluid to use is DEXRON®; do not use any other type. Drain the fluid from the transmission if the level is above the "F" on the dipstick, until the proper level is reached.

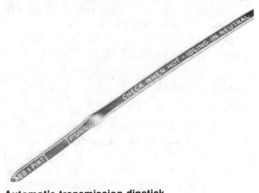

Automatic transmission dipstick

BRAKE MASTER CYLINDER

The fluid level in the brake master cylinder should be checked every six months. If necessary, it should be replenished to within ¼ in. of the top of the reservoir with brake fluid that meets DOT 3 or 4 requirements.

CAUTION: *Unlabeled fluid, or fluid with a lower boiling point, such as*

70R1, *should not be used, as a brake system failure may result.*

On cars equipped with front disc brakes, it is quite normal for the fluid level to fall slightly as the brake pads wear. However, cars with drum brakes on all four wheels should not have any noticeable drop in brake fluid level. Low fluid level may indicate the presence of a leak and the brake system should be checked carefully if this occurs.

Brake master cylinder

COOLANT

The coolant level should be checked at least once a month or when the temperature gauge registers "HOT" (H). Because the cooling system is under pressure, check the coolant level with the engine cold to prevent possible injury from a high-pressure stream of hot water.

The coolant level should be 1¼ in. below the filler neck of the radiator when the engine is cold. Replenish with 50% clean, nonalkaline water and 50% ethylene glycol antifreeze.

CAUTION: *Never add cold water to a hot engine. Damage to both the cooling system and the engine block could result.*

Some models are equipped with a closed cooling system, consisting of a tube running from the radiator to a thermal expansion tank. On these models, check the level of the coolant in the expansion tank. The main radiator cap should *only* be removed when cleaning or draining the cooling system. The level in the expansion tank should be at the "1 quart" mark, with engine running

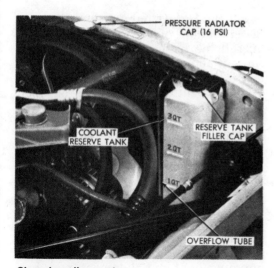

Closed cooling system

Anticipated Temperature Range	Viscosity Grade
Above 10° F	SAE 90
As low as −30° F	SAE 80
Below −30° F	SAE 75

If it is determined that the rear axle requires lubricant, add the appropriate amount and refit the filler plug. Lower the vehicle.

MANUAL STEERING LUBRICANT

Although regularly scheduled lubricant changes are not necessary, the lubricant level should be checked at six-month intervals. Remove the filler plug from the top rear of the steering gear

at idle and warmed up. Add coolant to the *expansion tank*, in order to bring it to this level.

CAUTION: *Do not add coolant to the main radiator on cars with a coolant recovery system.*

REAR AXLE LUBRICANT CHECKING

The rear axle lubricant level should be checked at every engine oil change. To check the level, jack the car and remove the axle filler plug. The correct lubricant level for each axle is indicated in the "Rear Axle Identification Chart." The manufacturer recommends that only multipurpose gear lubricant which meets the API GL-5 requirements to be used in conventional differentials. Use special limited slip fluid in Sure-Grip differentials. Gear lubricant viscosity depends on the anticipated ambient temperature.

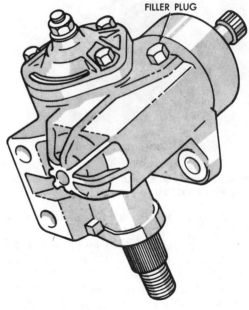

Manual steering gear

Rear Axle Identification Chart

Axle Size (ring gear diameter)	Filler Location	Cover Fastening	Capacity (pints)	Lubricant Level
7¼	Cover	9 bolts	2.0	Bottom of filler hole to ⅝ in. below
8¼	Carrier right side	10 bolts	4.4	From ⅛ in. below filler hole to ¼ in. below
8¾	Carrier right side	Welded	4.4	Maintain at bottom of filler hole
9¼ *	Cover	10 bolts	4.0	Bottom of filler hole to ½ in. below
9¾	Cover	10 bolts	5.5	Bottom of filler hole to ½ in. below

* Used 1968 and 1969 models only

housing and check to make sure that there is sufficient lubricant to cover the worm gear. If necessary, add SAE 90 multipurpose gear oil and then refit the filler plug.

POWER STEERING RESERVOIR

The fluid level in the power steering reservoir should be checked at every engine oil change. Locate the power steering pump in the engine compartment on the driver's side, below the cylinder head and adjacent to the block. Before removing the reservoir cover, wipe the outside of the cover and case so that no dirt can drop into the reservoir.

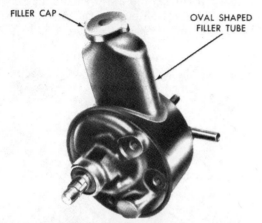

Power steering pump and reservoir (appearance may vary from year to year)

Check the fluid level when the engine is hot. Remove the reservoir cover. If it has a dipstick, the fluid level should be at the level indicated by the marks on the dipstick. If no dipstick is fitted, the correct fluid level is approximately ½–1 in. below the top of the filler neck. If necessary for either reservoir, add power steering fluid. *Do not use automatic transmission fluid.* Replace the reservoir cover.

BATTERY

The electrolyte level in the battery should be checked every few weeks. Remove the battery caps and check to see if the fluid level in the battery is up to the split rings in the bottom of the filler wells. If it is not, you can top it off with plain water unless you live in an area known to have hard water. If you do live in one of these areas, you should use distilled water.

Fill the battery only to the split rings in the bottom of the filler wells. If you overfill the battery, the electrolyte will run out and you will be left with a battery full of plain water.

Never light a match or use an open flame near the top of the battery, as the battery gives off *highly explosive* hydrogen gas.

Tire Care

In order to obtain the most uniform tread wear, your vehicle's tires should be rotated no later than every other oil change. Examine the tires at this time for unusual wear patterns. Erratic wear can be the result of improper inflation, improper front end alignment, poor balance or worn suspension parts. Do not mistake the tread wear indicators or bars molded into the bottom of the tread grooves as indicative of mechanical problems; when these ½ in. wide staggered bars appear, tire replacement is in order.

A decal showing the recommended tire pressure and size for your vehicle is located on the rear body pillar on the left front door. Proper pressure, balance, and size are important to both tread life and safety. It should be noted that the original balance of the tire and wheel are gradually lost as the tires wear; balance is also lost after a tire has been repaired.

If the vehicle is to be driven at speeds over 75 mph, increase the tire inflation pressure four psi over the normally recommended figure. However, in no case should inflation be increased over 32 psi for load range B tires. If the sticker on the car's body recommends a pressure differential between front and rear, it must be maintained.

When replacing your vehicle's tires, radial ply or wide belted types may be used, however, they must be used in sets of five and must never be mixed on a vehicle with tires of other than radial construction. Always select the recommended size radial.

If it becomes necessary to store your vehicle's tires, place them on their sidewalls on a flat surface. Permanent flat spotting can result if the tires are stored standing on their treads. If the tires are to be stored while mounted, inflation pressures should be reduced to 12–16

Capacities

Year	ENGINE No. Cyl Displacement (cu in.)	Engine Crankcase Add 1 Qt for New Filter	TRANSMISSION Pts to Refill After Draining			Drive Axle (pts)	Gasoline Tank (gals)	COOLING SYSTEM (qts)	
			Manual 3-Speed	4-Speed	Automatic			With Heater	With A/C
'68	6-225	4	6.5	——	15.5	2	24①	13	14
	8-318	4	6	——	19.5	4	24①	18	19
	8-383	4	6	9	15.5	4	24①	17	18
	8-440	4	——	9	15.5	4	24①	17	18
'69	6-225	4	6.5	——	15.5	2	24①	13	15
	8-318	4	6	——	15.5	4	24①	16	19
	8-383	4	6	7	18.5	4	24①	16	17
	8-383	4	——	——	15.5	4	24①	16	17
	8-440	4	——	——	18.5	4	24①	17	18
'70	6-225	4	4.75	——	17	2	24①	13	15
	8-318	4	4.75	——	16	4	24①	16	19
	8-383	4	4.75	——	19	4	24①	16	17
	8-383	4	4.75	——	16	4	24①	16	17
	8-440	4	——	——	19	4	24①	17	18
'71	6-225	4	4.75	——	17	4.5	23	13	13
	8-318	4	4.75	——	17	4.5	23	16	16.5
	8-360	4	4.75	——	16	4.5	23	15.5	15
	8-383	4	4.75	——	16③	4.5	23	14.5	15
	8-440	4	——	——	19	4	23	15.5	17
'72	8-318	4	——	——	17	4.5	23	16	17.5
	8-360	4	——	——	16.3	4.5	23	16	16
	8-400	4	——	——	19	4.5	23	14.5	15
	8-440	4	——	——	16.3	4.5	23	14.5	14.5
'73	8-318	4	——	——	17	4.5	23	16	17.5

Capacities (cont.)

Year	ENGINE No. Cyl (cu in.) Displacement	Engine Crankcase Add 1 Qt for New Filter	TRANSMISSION Pts to Refill After Draining			Drive Axle (pts)	Gasoline Tank (gals)	COOLING SYSTEM (qts)	
			Manual 3-Speed	4-Speed	Automatic			With Heater	With A/C
'73	8-360	4	—	—	16.3	4.5	23	15.5	16
	8-400	4	—	—	19	4.5	23	16	16
	8-440	4	—	—	19	4.5	23	15.5	15.5
'74	8-360	4	—	—	16.1	4.5	25②	16.5	16.5
	8-400	4	—	—	18.9④	4.5	25②	16.5	16.5
	8-440	4	—	—	16.1	4.5	25②	16	16
'75–'76	8-318	4	—	—	17⑥⑦	4.5	26.5②	17.5	17.5
	8-360	4	—	—	16.5⑥⑦	4.5	26.5②	16	16
	8-400	4	—	—	19⑦	4.5	26.5②	16.5	16.5
	8-440	4⑤	—	—	19⑦	4.5	26.5②	16	16

① Station wagon—23 gal
② Station wagon—24 gal
③ 2-bbl—19 pts
④ 16.1 pts with HP 400
⑤ 5 qts with HP 440
⑥ 11¾ in. torque converter—19 pts
⑦ With auxiliary cooler—20.5 pts

psi. Keep the tires away from water, oil, gasoline, electric motors (ozone), and heat.

← FRONT OF CAR

5 TIRE   4 TIRE

Bias-ply tire rotation

← FRONT OF CAR

5 TIRE   4 TIRE

Radial-ply tire rotation

Fuel Filter

All Plymouths are equipped with a throwaway filter which is located in the fuel line near the carburetor. On 1968-74 models the filter should be replaced every 2 years/24,000 miles, whichever occurs first. On 1976–76 models, replace the fuel filter every 15,000 miles.

To replace the filter, proceed in the following manner:

CAUTION: *Do not smoke while performing this procedure.*

1. Squeeze the clips securing the old filter assembly to the fuel line with pliers in order to open them.

2. Work the ends of both old rubber hoses off of the fuel line and lift out the filter with the hoses attached. Be careful not to drop gasoline on the manifolds if they are hot.

3. Using the clips supplied, install the pieces of rubber hose that come with the new filter over the necks on either side of it.

Typical fuel filter installation (arrow)

4. Install the filter hose assembly in the fuel line with clips, being sure that the arrow on the filter is pointing in the direction of fuel flow. Cut the hose to the correct length, if necessary.

5. Check to be sure that the fuel lines are not coming into contact with any other engine components.

6. Start the engine and check for leaks.

Battery Care

Maintain the battery electrolyte level, as outlined in the fluid level section, above.

If the terminals become corroded, clean them with a solution of baking soda mixed with water. Wash off the top of the battery with this solution and then rinse it off using clean, clear water.

CAUTION: *Be sure that the filler caps are on tight or the electrolyte in the battery may become contaminated.*

Use petroleum jelly or silicone lubricant to protect the battery terminals. Check to be sure that the cables are fastened securely at both ends. Also, be sure that the battery hold-down bracket nuts are secure and free of corrosion.

When installing a new battery, be sure that its amp/hour capacity is at least as high as that of the battery which was removed. Its physical size should be the same as that of the battery which it is replacing.

When hooking up the battery cables, be careful to observe proper polarity. The positive (hot) cable should be connected to the positive (+) terminal of the battery and the negative cable (ground) should be connected to the negative (−) terminal.

LUBRICATION

Oil and Fuel Recommendations

Use a good quality motor oil of a known brand, which carries the API classification SE. The proper viscosity of the oil to be used is determined by the chart below.

CAUTION: *Do not use unlabeled oil or a lower grade of oil which does not meet SE specifications.*

Change the oil at the intervals recommended in the lubrication chart below. If the vehicle is being used in severe service such as trailer towing, change the oil more frequently than recommended.

It is especially important that the oil be changed at the proper intervals in emission-controlled engines, as they run hotter than non-controlled engines, thus causing the oil to break down faster.

OIL VISCOSITY RECOMMENDATIONS

Multigrades

Where temperatures are consistently above 32° F	SAE 20W-40, SAE 10W-40 or SAE 10W-30
For year-long operation where temperatures occasionally drop to −10° F	SAE 10W-30 or SAE 10W-40

Single Grades

Where temperatures are consistently above 32° F	SAE 30
Where temperatures range between +32° F and −10° F	SAE 10W

Continuous high-speed operation or rapid acceleration requires heavier-than-normal lubricating oil. For the best protection under these conditions, use the heaviest oil that permits satisfactory cold starting. SAE 30 and 40 are recommended.

FUEL RECOMMENDATIONS

Prior to 1972 engines used in Plymouths varied widely as to the type fuel they used. Since all engines and gasolines vary in actual compression ratios and octane ratings, it is best to follow the recommendations in the owner's manual. If the manual is unavailable, consult the "General Engine Specifications" chart in Chapter 3 to determine your engine's compression ratio. As a rule of thumb, if compression ratio is 9.0:1 or higher, premium gasoline is required. Some experimenting with different grades of gasoline may be necessary to determine which works best in your car.

From 1972 on, all Plymouth engines are designed to run on regular gasoline having a research octane rating of 91. 1972 six-cylinder engines will run on unleaded or low-lead fuels, while V8s require leaded fuel with every fourth fill up. On 1973–74 models the use of induction-hardened valve seats allows V8s to run only on no-lead or low-leaded fuels,

as well; although their use is not mandatory.

Catalytic converts are used on most 1975–76 models. When a car is equipped with a catalytic converter, it requires the use of unleaded fuel *only*. These cars have labels, both near the fuel filler and on the instrument panel, which read "UNLEADED GASOLINE ONLY". To further prevent the use of leaded fuel, the filler neck is smaller than previously used and has a restriction to allow only the new, small unleaded fuel pump nozzle to fit into it. Plymouth also requires that fuel system cleaning agents, which are added to the fuel tank or carburetor, not be used, as they can contaminate the catalyst.

Cars which do not have catalytic converters follow the same fuel recommendations as 1973–74 models.

If the engine pings, knocks, or diesels, either the fuel grade is too low or the timing is out of adjustment. Add gasoline of a higher octane and check the timing, as soon as possible.

CAUTION: *Pinging, knocking, or dieseling can rapidly damage the engine, the problem should be cured as quickly as possible.*

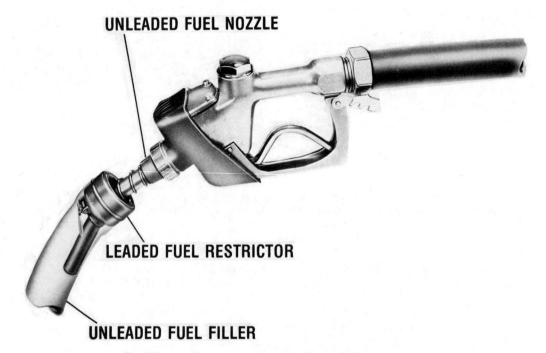

UNLEADED FUEL NOZZLE

LEADED FUEL RESTRICTOR

UNLEADED FUEL FILLER

Fuel filler used on cars equipped with catalytic converters

Oil Changes

ENGINE

The oil should be changed at the following intervals, under normal operating conditions:

1968–74 models—3 months/4,000 miles
1975–76 models—6 months/5,000 miles

If the car is frequently operated under any of the following conditions, the engine oil should be changed at least twice as often as above:

1. Driving under dusty conditions—
2. Trailer towing—
3. Long periods of idling—
4. Short trip (less than 10 miles) in cold weather (10° F or less)—
5. High-speed driving (70 mph +) in hot (90° F +) weather.

To change the oil, proceed in the following manner:

1. Warm the oil by running the engine for a short period of time; this will make the oil flow more freely from the oil pan.
2. Park on a level surface and put on the parking brake. Stop the engine. Remove the oil filler cap from the top of the valve cover.
3. Place a pan of adequate capacity below the drain plug.

NOTE: *If the crankcase holds four quarts, a two-quart milk container will not be suitable. A large flat pan makes a good container to catch oil.*

4. Use a wrench of the proper size (not pliers) to remove the drain plug. Loosen the drain plug while maintaining a slight upward force on it to keep the oil from running out around it. Allow the oil to fully drain into the container under the drain hole.
5. Remove the container used to catch the oil and wipe any excess oil from the area around the hole.
6. Install the drain plug, complete with its gasket if used. Be sure that the plug is tight enough that the oil does not leak out, but not tight enough to strip the threads.

NOTE: *Replace the drain plug gasket if used with a new one.*

7. Add clean, new oil of the proper grade and viscosity through the oil filler on the top of the valve cover. Be sure that the oil level registers near the "FULL" mark on the dipstick.

8. Start the engine and allow it to idle for a few minutes. Stop the engine. Remove the dipstick and check the oil level. Add oil, if necessary, to bring the level up to the full mark. Do not overfill.
9. Remember to replace the oil filler cap. Check for oil leaks.

MANUAL TRANSMISSION

Cars used in normal service do not require a transmission oil change at regular specified intervals. However, if the car is used for commercial operation (police, taxi, etc.) or for trailer towing, the transmission lubricant should be replaced every 3 years/36,000 miles, whichever occurs first.

To change the transmission oil, proceed as follows:

1. Park the car on a level surface and put on the parking brake.
2. Remove the oil filler (upper) plug.
3. Place a container, of a large enough capacity to catch all of the oil, under the drain (lower) plug. Use the proper size wrench to loosen the drain plug slowly, while maintaining a slight upward force to keep the oil from running out. Once the plug is removed, allow all of the oil to drain from the transmission.
4. Install the drain plug.
5. Fill the transmission to capacity (see "Capacities" chart) with DEXRON® automatic transmission fluid.

NOTE: *If the gears rattle at idle or during acceleration, drain the automatic transmission fluid and fill the transmission with multipurpose gear lubricant SAE 90 or 140.*

6. Install the filler plug when finished.

AUTOMATIC TRANSMISSION

Under normal operating conditions, automatic transmission fluid changes are not necessary and not recommended. However, if the car is operated under one of the following conditions, the transmission fluid and filter should be changed and the bands adjusted, as well, every 24,000 miles on 1968–74 models or every 25,000 miles on 1975–76 models:

1. Over 50% operation in heavy traffic when the ambient temperature is 90° F or higher.
2. Commercial operation

3. Trailer towing

See Chapter 6, "Clutch and Transmission" for the proper procedures for DEXRON® fluid change, filter change, and band adjustment.

REAR AXLE

For passenger cars used in normal service, it is unnecessary to to change the lubricant in the rear axle. However, if one of the following conditions applies, the lubricant should be changed:

1. If the lubricant has become contaminated with water, it should be changed as soon as possible to prevent damage to the axle. Contamination occurs when the axle is submerged in water deep enough to cover its vent; for example, when using a boat launching ramp.

2. If the car is used for commercial service (police, taxi, etc.) or for trailer towing, the lubricant should be changed every 36,000 miles on 1968–74 models or every 35,000 miles on 1975–76 models.

3. The only other time it should be necessary to change the lubricant, is to provide the proper viscosity to the anticipated ambient temperature conditions.

For the type of lubricant to use, as well as for correct viscosity, see the rear axle section of "Fluid Level Checks".

In order to change the lubricant, proceed as follows:

1. Park the car on a level surface and set the parking brake.

2. Crawl underneath the rear of the car and remove the filler plug.

3. If the axle has a drain (lower) plug, place a container of adequate capacity underneath it. Remove the plug and allow the lubricant to drain into the container.

4. On axles without drain plugs, use a suction gun to remove the lubricant through the filler plug hole. Drain it into a suitable container.

5. Install the drain plug, if used, and tighten it so that it won't leak; but do not overtighten it.

6. Refill the axle with the proper grade and viscosity of axle lubricant. Be sure that the lubricant reaches the fill level specified in the "Rear Axle Identification" chart.

7. Install the filler plug and check for leaks.

Removing the rear axle lubricant with a suction gun

Oil Filter Changes

Plymouth recommends that the oil filter be changed at the first oil change and every second oil change after that.

On 1972 and earlier models, the "long" type of oil filter must be used; while on 1973–76 models the newer, "short" type of oil filter should be used.

NOTE: *On 1973–76 models using 400 or 440 cu in. engines which are equipped with an air injection pump and/or power steering, the short filter must be used. On all other 1973–76 models the use of the long filter is acceptable.*

To replace the filter, proceed as follows:

1. Drain the engine oil as outlined above. Place a container under the oil filter to catch any excess oil.

2. Use a spin-off (band) wrench to remove the filter unit. Turn the filter counterclockwise in order to remove it.

3. Wipe off the filter bracket with a clean rag.

4. Install the new filter and gasket,

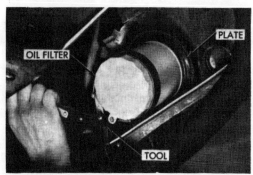

Use a band wrench to remove the oil filter

after first lubricating the gasket with clean engine oil.

CAUTION: *Do not use the wrench to tighten the filter. Tighten it by hand.*

5. Add engine oil as outlined in the appropriate section above. Check for leaks.

Chassis Greasing

All Chrysler vehicles are equipped with nine ball joints or end fittings in the front suspension and steering system. They should be inspected for wear and damage twice yearly, and lubricated every 18,000 miles, for vehicles used in severe service, or 36,000 (35,000 miles—1975–76 miles for vehicles used in normal service. For information on checking ball joint wear, please refer to the "Front Suspension" section. For lubrication procedures, see below:

1. With a clean rag, wipe all road dirt from the ball joint seal or end fitting.

2. Take the threaded plug from the ball joint and install a grease fitting (if applicable).

3. Fill the joint or fitting with lubricant. The joint is filled when grease freely flows from the bleed area at the base of the seal or fitting, or the seal begins to balloon.

NOTE: *It is better to use a hand-operated grease gun, in order to avoid seal damage; however, if a high-pressure gun is used, use care not to over-fill the seal.*

Throttle Linkage Lubrication

The throttle linkage should be lubricated at least once a year with multipur-

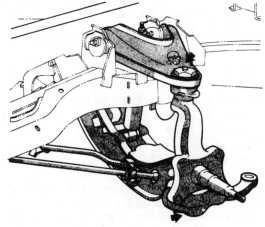

Front ball joint lubrication points

pose grease, NGLI grade 2EP. Be sure to wipe the linkage clean in order to remove all accumulated grease and grime.

CAUTION: *Do not lubricate the throttle cable or linkage ball joints.*

Body Lubrication

Aside from the hood latch, which should be lubricated every six months, there is no required schedule for body lubrication. However, it is a good idea to lubricate all of the parts listed below at the same time that the hood latch is being lubricated. If any parts start to squeak or bind, they should be lubricated as soon as possible.

Before lubricating any body part, wipe it clean to remove accumulated old lubricant and grime. After applying the required lubricant, wipe away any excess; this is particularly important in areas which could come into contact with clothing.

In cold areas, particular attention should be paid to door and trunk lock cylinder lubrication to help prevent them from freezing.

Lubricate the pivoting and sliding surfaces of the following items with the type of lubricant specified:

1. Engine oil:
   - Door hinges
   - Hood hinges
   - Tailgate hinges

NOTE: *Do not lubricate the roller surfaces of the hinge arms and roller on the front or rear door lower hinges.*

2. Multipurpose grease, NGLI grade 2EP:
   - Hood latch and related items
   - Fuel filler door hinges
   - Concealed headlight door shaft
   - Tailgate torsion bar and "antislam" assembly

3. Smooth, white body lubricant, NGLI grade 1:
   - Hood hinge cam and roller
   - Trunk lid torsion bar
   - Lock cylinders (apply to key and insert)
   - Parking brake assembly
   - Door latch
   - Trunk latch(es)
   - Tailgate latch

4. Stainless wax stick lubricant:
   - Door latch striker plate/bolt

- Tailgate latch striker plate/bolt
- Ashtray slides
5. Spray silicone lubricant:
   - Weather stripping
   - Rubber hood stops

Points Not Requiring Lubrication

The following items should *not* be lubricated, since they are already lubricated; lubrication will cause their failure or will cause them to work incorrectly:

- All rubber bushings;

CAUTION: *Be extremely careful not to lubricate rubber bushings, the lubricants will cause them to fail and destroy their frictional characteristics as well.*

- Clutch adjustment rod ends
- Clutch pedal pushrod ends
- Clutch throwout bearing
- Drive belts and pulleys
- Rear springs
- Rear wheel bearings
- Starter and alternator bearings
- Throttle cable
- Throttle linkage ball joints
- Automatic transmission selector linkage
- Control arm bushings
- Water pump bearings
- Air cleaner element
- Accelerator pedal pivots

Front Wheel Bearings

The front wheel bearings should be repacked and adjusted at the following intervals:

- 1968–73—every year/12,000 miles
- 1974—every 2 years/24,000 miles
- 1975–76—every 25,000 miles

NOTE: *Adjust and repack the front wheel bearings whenever the front brake disc or drum is serviced.*

See the appropriate section in Chapter 9 below for the correct front wheel bearing repacking and adjustment procedure.

PUSHING, TOWING, AND JUMP STARTING

Push-start the engine when it will not turn over; do not attempt to start the car by towing it.

CAUTION: *If the car is tow-started, it may run into the back of the towing vehicle when it starts.*

To push-start the car, turn the ignition switch to "on." Fully depress the clutch pedal and shift into second or third gear. When the car reaches 10 mph, let the clutch pedal up slowly until the engine catches.

NOTE: *It is impossible to push-start models equipped with the TorqueFlite automatic transmission.*

The following precautions should be observed whenever towing a vehicle:

1. Always place the transmission in Neutral (N) and release the parking brake.

2. Models equiped with TorqueFlite may be towed with the transmission in Neutral (N) but only for distances of 15 miles at speeds not exceeding 30 mph. If the transmission is inoperative or the car must be towed for a distance of more than 15 miles, either tow the car with the rear wheels off the ground or disconnect the driveshaft.

3. If the rear axle is defective, the car must be towed with its rear wheels off the ground.

4. Always turn the steering column lock to the "OFF" (Unlocked) position, if the car is being towed with the front wheels down.

CAUTION: *The steering column lock is not designed to hold the straight-ahead position. Therefore, if the car is towed with its front end down, either use a special steering wheel clamp or place a dolly under the wheels.*

5. When lifting the back end of the car, do not allow any of the towing equipment to contact the gas tank or the brake lines.

6. Remove the exhaust extensions or any other equipment, except bumper guards, which interfere with the towing sling.

7. Cars must *not* be towed by hooking on to any of the following: rear spring shackles, front brake struts, stabilizer bar, or bumper hydraulic energy absorbers (1974–76).

8. A separate safety chain system should be used when towing the car.

9. Use lumber and/or padding to protect the car. See the illustrations for the proper towing sling installation.

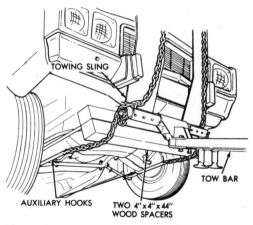

Typical front end towing—1968–73

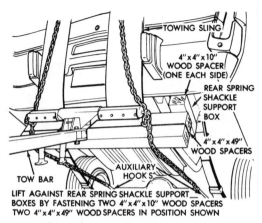

Typical rear end towing—1968–73

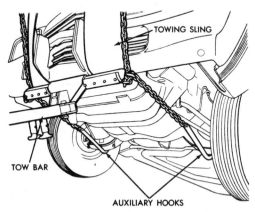

Typical front end towing—1974–76

NOTE: *If two pieces of lumber must be used, be sure that they are securely nailed together and wired to the tow chains.*

When jump starting a car, be sure to observe the proper polarity of the battery connections.

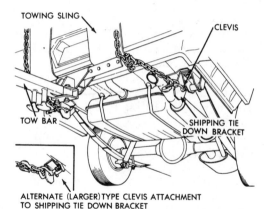

Typical rear end towing—1974–76 sedans

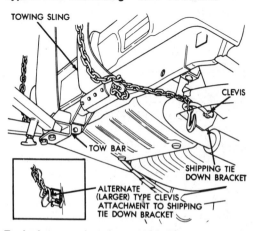

Typical rear end towing—1974–76 wagons

CAUTION: *Always hook up the positive (+) terminal of the booster battery to the positive terminal of the discharged battery and the negative terminal (−) to a good ground.*

If the battery terminals are unmarked, the correct polarity of each battery may be determined by examining the battery cables. Because Plymouths are negative (−) ground, the negative (ground) cable will run to the chassis and the positive (hot) cable will run to the starter motor. A 12 volt fully charged battery should be used for jump starting.

To jump start the car, proceed in the following manner:

1. Put the transmission in Park (P) or Neutral (N) and set the parking brake. Make sure that all electrical loads are turned off (lights, wipers, etc.).

2. Remove the vent caps from both batteries. Cover the opened vents of both batteries with a clean cloth. These two steps help reduce the hazard of an explosion.

3. Connect the positive (+) terminals of both batteries first. Be sure that the cars are not touching or the ground circuit may be completed accidentally.

4. Connect the negative (−) terminal of the booster battery to a suitable ground (engine lifting bracket, alternator bracket, etc.) on the engine of the car with the dead battery. Do not connect the negative jumper cable to the ground post of the dead battery.

5. Start the car's engine in the usual manner.

6. Remove the jumper cables in *exactly* the reverse order used to hook them up.

JACKING AND HOISTING

Specific instructions for the use of the jack and the proper jacking points are found on stickers which are located in the luggage compartment.

There are, however, some general precautions to observe when jacking the automobile:

1. Never climb underneath the car when it is supported only by the jack. Always use jackstands as an additional means of support.

2. Always keep the car on level ground when jacking it. Otherwise, the car may roll or fall off the jack.

3. When raising the *front* of the car, set the parking brake and block the rear wheels.

4. When raising the *rear* of the car, securely block the front wheels.

If the car is being raised with a frame contact hoist, be sure that the proper adapters are available to support the car in the locations illustrated.

Regular hydraulic hoists may be used if they will make secure contact with the lower control arms and the rear axle housing.

Floor jacks may be used *only* under the rear axle housing or the lower control arms.

The following precautions should be observed whenever the car is being hoisted:

1. Never place the hoist or jack lifting pads under the sheet metal parts of the body. Permanent damage to the body may result when the car is raised.

2. Never raise only one side of the car by placing a floor jack midway between the front and rear wheels. Permanent damage to the frame may result when the side is lifted.

3. When using a floor jack to raise the rear of the car, do not allow the lifting plate fingers to contact the axle cover plate.

4. When using a bumper jack to raise the car, be sure to use one which will engage the notches in the bumper.

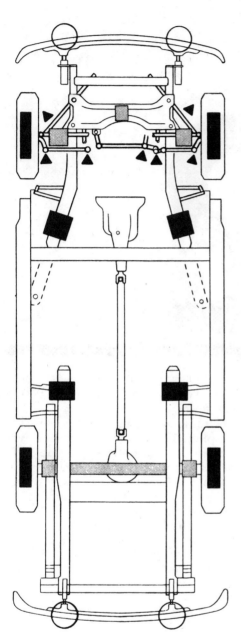

LUBRICATION
▲ SUSPENSION AND STEERING LINKAGE BALL JOINTS
LIFTING
■ FRAME CONTACT OR DRIVE ON HOIST
▨ FLOOR JACK OR HOIST
○ BUMPER JACK (AT BUMPER SLOT ONLY)

Lifting, jacking, hoisting, and lubrication points

Tune-Up

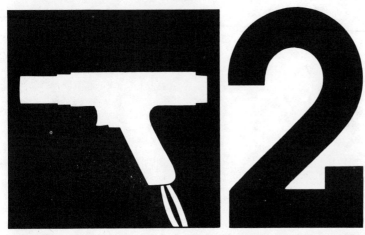

TUNE-UP PROCEDURES

This section gives specific procedures on how to tune-up your Plymouth. It is intended to be as complete and as basic as possible. There are also Engine troubleshooting sections given in Chapter 11. Perhaps the best procedure to follow would be to read both sections before starting your tune-up.

Spark Plugs

The job of the spark plug is to ignite the air/fuel mixture in the cylinder as the piston approaches the top of the compression stroke. The ignited mixture then expands and forces the piston down on the power stroke. This turns the crankshaft which then turns the remainder of the drive train. If the transmission is in gear and the brakes are not applied, the vehicle will move.

The average life of a conventional spark plug is approximately 12,000 miles. This is, however, dependent on the mechanical condition of the engine, the type of fuel that is used, and the type of driving conditions under which the car is used. For some people, spark plugs will

last 5,000 miles and for others, 15,000 or 20,000 miles.

Electronic ignition generally increases spark plug life. Electronic ignition was first offered as an option on some 1972 models and became standard on all Plymouths in 1973. Because of this, the recommended spark plug replacement interval was increased to 18,000 miles for models with electronic ignition.

The use of unleaded fuel also increases spark plug life. Most 1975–76 Plymouths are equipped with catalytic converters which require the use of unleaded fuel exclusively. The spark plug changing interval has been modified accordingly:

• 1975–76 models without converters —every 15,000 miles
• 1975–76 models with converters— every 30,000 miles

Many 1975–76 Plymouths equipped with 400 and 440 cu in. V8 engines use platinum-tipped "long life" spark plugs. The recommended replacement intervals for these are as follows:

• 1975–76 models without converters —every 30,000 miles
• 1975–76 models with converters— every 50,000 miles

Of course the same conditions which

Tune-Up Specifications

When analyzing compression test results, look for uniformity among cylinders rather than specific pressures.

Year	ENGINE No. Cyl Displacement (cu in.)	Hp	SPARK PLUGS Type	Gap (in.)	DISTRIBUTOR Point Dwell (deg)	Point Gap (in.)	IGNITION TIMING (deg) ▲ Man Trans •	Auto Trans	Valves Intake Opens (deg) ■	Fuel Pump Pressure (psi)	IDLE SPEED (rpm) ▲ Man Trans •	Auto Trans
'68	6-225	145	N-14Y	.035	40-45	.020	5B(TDC)	5B(TDC)	10	3½-5	550(650)	550(650)
	8-318	230	N-14Y	.035	28-33	.017	5B(5A)	10B(2½A)	10	5-7	550(650)	500(600)
	8-383	290	J-14Y	.035	28-33	.017	TDC	7½B	18	3½-7	650	600
	8-383	330	J-11Y	.035	28-33	.017	TDC	5B	18	3½-5	650	650
	8-440	350	J-13Y	.035	28-33	.017	—	7½B	18	3½-5	—	600
	8-440	375	J-11Y	.035	28-33①	.017	TDC	5B	21	6-7½	650	650
'69	6-225	145	N-11Y	.035	42-47	.020	TDC	TDC	10	3½-5	700	650
	8-318	230	N-14Y	.035	30-35	.017	TDC	TDC	10	5-7	700	650
	8-383	290	J-14Y	.035	30-35	.017	TDC	7½B	18	3½-5	700	600
	8-383	330	J-11Y	.035	30-35	.017	TDC	5B	18	3½-5	700	650
	8-440	350	J-13Y	.035	28-33	.017	—	7½B	18	3½-5	—	600
	8-440	375	J-11Y	.035	②	.017	TDC	5B	21	6-7½	700	650
'70	6-225	145	N-14Y	.035	41-46	.020	TDC	TDC	10	3½-5	700	650
	8-318	230	N-14Y	.035	30-34	.017	TDC	TDC	10	5-7	750	700
	8-383	290	J-14Y	.035	28-32	.018	TDC	2½B	18	3½-5	750	650
	8-383	330	J-11Y	.035	28-32	.018	TDC	2½B	18	3½-5	750	750
	8-440	350	J-13Y	.035	28-33	.018	—	12½B	18	3½-5	—	650
	8-440	390	J-11Y	.035	27-32①	.017	5B	5B	21	6-7½	900	900
'71	6-225	145	N-14Y	.035	41-46	.020	TDC(2½B)	TDC(2½B)	16	3½-5	750	750
	8-318	230	N-14Y	.035	30-34	.017	TDC	TDC	10	5-7	750	700

Tune-Up Specifications (cont.)

When analyzing compression test results, look for uniformity among cylinders rather than specific pressures.

Year	ENGINE No. Cyl Displacement (cu in.)	Hp	SPARK PLUGS Type	Gap (in.)	DISTRIBUTOR Point Dwell (deg)	Point Gap (in.)	IGNITION TIMING (deg) Man Trans ●	Auto Trans ●	Valves Intake Opens (deg) ■	Fuel Pump Pressure (psi)	IDLE SPEED (rpm) ▲ Man Trans ●	Auto Trans ▲
	8-360	255	N-13Y	.035	30-34	.017	2½B	2½B	16	3½-5	750	700
	8-383	275	J-14Y	.035	28-32	.018	TDC	2½B	18	3½-5	750	700
	8-383	300	J-11Y	.035	28-32	.018	TDC	2½B	21	3½-5	900	800
	8-440	335	J-13Y	.035	28-32	.018	—	5B	18	3½-5	—	750
	8-440	370	J-11Y	.035	28-32	.018	TDC	2½B	21	3½-5	900	800
'72	8-318	150	N-13Y	.035	30-34	.017	—	TDC	10	5-7	—	750(700)
	8-360	175	N-13Y	.035	30-34	.017	—	TDC	16	5-7	—	750
	8-400	190	J-13Y	.035	28-32	.018	—	5B③	18	3½-5	—	700
	8-440	225	J-11Y	.035	28-32	.018	—	10B	18	3½-5	—	750(700)
'73	8-318	150	N-13Y	.035	Electronic	—	—	TDC	10	6-7½	—	700
	8-360	170	N-13Y	.035	Electronic	—	—	TDC	16	6-7½	—	750
	8-400	185	J-13Y	.035	Electronic	—	—	10B	18	4-5½	—	700
	8-440	220	J-11Y	.035	Electronic	—	—	10B	18	4-5½	—	700
'74	8-360	180	N-12Y	.035	Electronic	—	—	5B	16	5-7½	—	750
	8-400	185	J-13Y	.035	Electronic	—	—	5B	18	4-5½	—	750
	8-400HP	260	J-13Y	.035	Electronic	—	—	5B	18	4-5½	—	900(750)
	8-440HP	275	J-11Y	.035	Electronic	—	—	10B	18	7-8.2	—	750
'75	8-318	150	N-13Y	.035	Electronic	—	—	2B	10	5-7	—	750
	8-360	All	N-12Y	.035	Electronic	—	—	6B	18	5-7	—	750
	8-400	175	J-13Y	.035	Electronic	—	—	10B	18	4-5½	—	750

Engine										
8-400	195	J-13Y	.035	Electronic	—	8B	18	4-5½	—	750
8-440	215	RY-87P	.040	Electronic	—	8B	18	4-5½	—	750
8-318 ①	150	RN-12Y	.035	Electronic	—	2B	10	5-7	—	750
8-318 ⑤	150	RN-12Y	.035	Electronic	—	2A	10	5-7	—	750
8-360	170	RN-12Y	.035	Electronic	—	6B	18	5-7	—	700
8-360 Cal	175	RN-12Y	.035	Electronic	—	6B	18	5-7	—	750
8-400	175	RJ-13Y	.035	Electronic	—	10B	18	5-7	—	700
8-400 Cal	185	RJ-13Y	.035	Electronic	—	8B	18	5-7	—	750
8-400 LB	210	RJ-13Y	.035	Electronic	—	10B	18	5-7	—	720
8-400 HP	240	RJ-86P	.035	Electronic	—	6B	18	5-7	—	750
8-440	205	RJ-13Y	.035	Electronic	—	8B	18	5-7	—	750
8-440 HP	255	RJ-11Y	.035	Electronic	—	10B	18	6-7.5	—	750

'76

▲ See text for procedure
■ Before Top Dead Center
• Figure in parentheses indicates California engine
① Both sets 37°-40°
② Automatic transmission 30°-50°
Manual 27°-32°, both sets 37°-40°
③ Non-California cars built after Feb. 2, 7½B
④ With catalyst
⑤ With air pump only

A After Top Dead Center
B Before Top Dead Center
TDC Top Dead Center
HP High Performance or Police Interceptor
LB Lean burn engine
Cal California only

Mechanical Valve Lifter Clearance

Year	Engine	Intake (Hot) In.	Exhaust (Hot) In.
1968-1971	All six cylinder	.010	.020

affect spark plug life normally, still apply. Thus, you might find that you have to replace plugs sooner, or even at less frequent intervals, than recommended.

The electrode end of the spark plug (the end that goes into the cylinder) is also a very good indicator of the mechanical condition of your engine. If a spark plug should foul and begin to misfire, you will have to find the condition that caused the plug to foul and correct it. It is also a good idea to occasionally give all the plugs the once-over to get an idea how the inside of your engine is doing. A small amount of deposit on a spark plug, after it has been in use for any period of time should be considered normal. But a black liquid deposit on the plugs indicates oil fouling. You should schedule a few free Saturday afternoons to find the source of it. Because the combustion chamber is supposed to be sealed from the rest of the engine, oil on the spark plug means your engine is hemorrhaging. Don't worry, though, this only happens once in a while.

REMOVAL

1. If the spark plug wires are not numbered as to their cylinder, place a piece of masking tape on each wire and number it.

2. Grasp each wire by the rubber boot at the end. Pull the wires from the spark plugs. If the boots stick to the plugs, remove them with a twisting motion. Do not attempt to remove the spark plug wires from the plugs by pulling on the wire itself as this will damage the spark plug wires.

3. Using a $13/16$ in. spark plug socket, loosen each plug several turns.

4. If compressed air is available, blow off the area around each spark plug hole. Otherwise, use a rag or other suitable material and remove any loose particles from around each plug hole. In either case, make sure that foreign matter is not allowed to enter the cylinders.

5. Unscrew the plugs the rest of the way and remove them from the engine.

INSPECTION

Compare the condition of the spark plugs to the plugs shown in the "Color"

section of chapter 4. It should be remembered that any type of deposit will decrease the efficiency of the plug. If the plugs are not to be replaced, they should be thoroughly cleaned before installation. If the electrode ends of the plugs are not worn or damaged and if they are to be reused, wipe off the porcelain insulator on each plug and check for cracks or breaks. If either condition exists, the plug must be replaced.

If the plugs are judged reusable, have them cleaned on a plug cleaning machine (found in most service stations) or remove the deposits with a stiff wire brush.

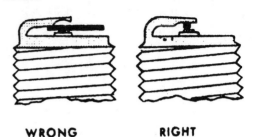

WRONG   **RIGHT**

Check the spark plug gap with a round wire gauge

Check the plug gap on both new and used plugs before installing them in the engine. The ground electrode must be parallel to the center electrode and the specified size wire gauge should pass through the opening with a slight drag. If the center or ground electrode has worn unevenly, level them off with a file. If the air gap between the two electrodes is not correct, open or close the ground electrode, with the proper tool, to bring it to specifications. Such a tool is usually provided with a gap gauge.

INSTALLATION

1. Insert the plugs into the engine and tighten them hand tight. Do not cross-thread the plugs.

When replacing spark plugs on a six-cylinder engine, do not use the metallic gaskets supplied with the plug. In addition, it is recommended, for engines utilizing spark plug tubes, that the inner rubber tube gasket be replaced.

2. Tighten the plugs to 30 ft lbs.

3. Install the spark plug wires on their respective plugs, making sure each wire is firmly connected.

Breaker Points and Condenser—1968-72

The points and condenser function as a circuit breaker for the primary circuit of the ignition system. The ignition coil must boost the 12 volts (V) of electrical pressure supplied to it by the battery to about 20,000 V in order to fire the spark plugs. To do this, the coil depends on the points and condenser for assistance.

The coil has a primary and a secondary circuit. When the ignition key is turned to the "on" position, the battery supplies voltage to the primary side of the coil which passes the voltage on to the points. The points are connected to ground to complete the primary circuit. As the cam in the distributor turns, the points open and the primary circuit collapses. The magnetic force in the primary circuit of the coil cuts through the secondary circuit and increases the voltage in the secondary circuit to a level that is sufficient to fire the spark plugs. When the points open, the electrical charge contained in the primary circuit jumps the gap that is created between the two open contacts of the points. If this electrical charge was not transferred elsewhere, the material on the contacts of the points would melt and that all-important gap between the contacts would start to change. If this gap is not maintained, the points will not break the primary circuit. If the primary circuit is not broken, the secondary circuit will not have enough voltage to fire the spark plugs. Enter the condenser.

The function of the condenser is to absorb the excessive voltage from the points when they open and thus prevent the points from becoming pitted or burned.

If you have ever wondered why it is necessary to tune-up your engine occasionally, consider the fact that the ignition system must complete the above cycle each time a spark plug fires.

There are two ways to check breaker point gap: with a feeler gauge or with a dwell meter. Either way you set the points, you are adjusting the amount of time (in degrees of distributor rotation) that the points will remain open. If you adjust the points with a feeler gauge, you are setting the maximum amount the

points will open when the rubbing block on the points is on a high point of the distributor cam. When you adjust the points with a dwell meter, you are measuring the number of degrees (of distributor cam rotation) that the points will remain closed before they start to open as a high point of the distributor cam approaches the rubbing block of the points.

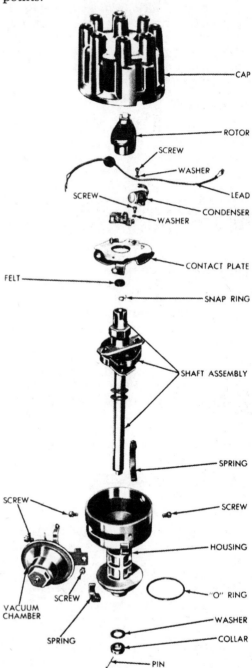

Typical distributor assembly

If you still do not understand how the points function, take a friend, go outside, and remove the distributor cap from your engine. Have your friend operate the starter (make sure the transmission is not in gear) as you look at the exposed parts of the distributor.

There are two rules that should always be followed when adjusting or replacing points. *The points and condenser are a matched set; never replace one without replacing the other. If you change the point gap or dwell of the engine, you also change the ignition timing. Therefore, if you adjust the points, you must also adjust the timing.*

INSPECTION

1. Disconnect the high-tension wire from the coil and unsnap the distributor cap retaining clips.

2. Remove the distributor cap from the distributor and position it out of the way.

3. Remove the rotor from the distributor shaft by pulling it straight up. Examine the condition of the rotor, if it is cracked or the metallic tip is excessively burned, it should be replaced.

4. Place a screwdriver against the breaker points and pry them open. Examine the condition of the contact points. If they are excessively worn, burned, or pitted they should be replaced.

5. If the points are in good condition and not in need of replacement, proceed to the breaker point adjustment section which follows the breaker point replacement procedure. If the points need to be replaced, use the following point removal procedure.

REMOVAL AND INSTALLATION, ADJUSTMENT

Single-Point Distributor

Use the procedure described below to remove, install, and gap a single-contact point set.

1. Pull back the spring clips and lift off the distributor cap. Pull off the rotor.

2. Loosen the terminal screw nut and remove the primary and condenser leads.

3. Remove the stationary contact lockscrew and remove the contact point set.

4. Remove the condenser retaining screw and lift out the condenser.

5. Install the new condenser and tighten its retaining screw.

6. Install the new point set but do not fully tighten its lockscrews.

7. Connect the condenser and primary leads.

8. If necessary, align the contacts by bending the stationary contact bracket only. *Never bend the movable contact arm to correct alignment.*

9. Attach a remote starter switch to the electrical system according to the switch manufacturer's instructions. Use this switch to crank the engine, rotating the distributor cam until the rubbing block of the movable contact arm rests on a peak of the cam lobe. If no remote starter switch is available, the same result may be obtained by gently tapping the ignition key to allow the starter to rotate the engine only a small amount. Adjust the points as above.

10. Insert the proper thickness feeler gauge between the contact points. If necessary, increase or decrease the gap by inserting a screwdriver in the "V" notch of the stationary contact base and using the screwdriver to move the stationary contact.

11. Tighten the lockscrew and re-

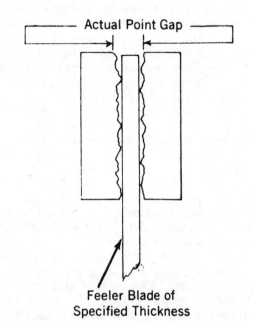

Feeler Blade of Specified Thickness

Using a feeler blade (gauge) is a much less accurate method of setting the points than with a dwell meter

check the gap setting. Reset if necessary.

12. Install the new rotor and refit the distributor cap. Check the point dwell.

Dual-Point Distributor

Removal and installation of dual contact points is the same as for single point set. However, adjustment of dual points is slightly different because one set of contacts must be blocked open with a clean insulator while the opposite points set is adjusted to specifications, using the single-point set adjustment procedure. When adjusted correctly, tighten the lockscrew. Block open this contact set and adjust the other set in the same manner as for the first. Check the point dwell. If the contacts have been installed and adjusted correctly, the dwell angle should be as specified for both contact sets.

Dwell Angle

1. Run the engine and allow it to reach normal operating temperature. Stop the engine.

2. Connect a dwell meter (or dwell/tachometer) to the engine according to the manufacturer's instructions. One wire, usually the black one, goes to a ground, such as a an engine lifting bracket, sheet metal screw, etc. The other wire usually the red one, should be clipped to the distributor terminal on the coil.

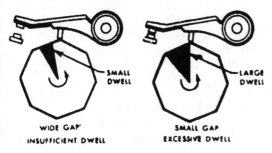

WIDE GAP   SMALL GAP
INSUFFICIENT DWELL   EXCESSIVE DWELL

Dwell is a function of point gap

NOTE: *The correct terminal may be identified by the small wire which runs from it to the distributor base. The large cable which runs to the distributor cap is the high tension lead and should not be disturbed.*

3. If the dwell meter has a set line on it, use a small screwdriver to adjust the needle so that it aligns with the line.

4. Start the engine. Be sure that the transmission is in Neutral (N) or Park (P) and that the parking brake is on.

CAUTION: *Be careful not get wires, tools, hands, clothing, etc., near the engine fan.*

5. Observe the reading on the dwell meter. If the reading is within specifications, stop the engine and disconnect the dwell meter.

6. If the reading is above specifications, the point gap is too small; if the reading is below specifications, the point gap is too large.

7. Turn off the engine and remove the distributor cap. Adjust the point gap by loosening the lockscrew, inserting a screw driver in the "V" notch of the stationary contact base, and moving the stationary contact as required.

8. Tighten the lockscrew, install the cap, and recheck the dwell. When the dwell is correct, turn off the engine and disconnect the dwell meter.

NOTE: *See "Dual-Point Distributor", for the correct dwell adjustment of these units.*

9. Check and adjust the ignition timing after completing the dwell adjustment.

Electronic Ignition—1972–76

This system consists of a special pulse-sending distributor, an electronic control unit, a two-element ballast resistor, and a special ignition coil.

The distributor does not contain breaker points or a condenser, these parts being replaced by a distributor reluctor and a pick-up unit.

The ignition primary circuit is connected from the battery, through the ignition switch, through the primary side of the ignition coil, to the control unit where it is grounded. The secondary circuit is the same as in conventional ignition systems: the secondary side of the coil, the coil wire to the distributor, the rotor, the spark plug wires, and the spark plugs.

The magnetic pulse distributor is also connected to the control unit. As the distributor shaft rotates, the distributor reluctor turns past the pickup unit. As the reluctor turns past the pick-up unit, each of the six or eight teeth on the reluctor pass near the pick-up unit once during each distributor revolution (two crankshaft revolutions since the distributor runs at one-half crankshaft speed). As the reluctor teeth move close to the pick-up unit, the magnetic rotating reluctor induces voltage into the magnetic pick-up unit. This voltage pulse is sent to the ignition control unit from the magnetic pick-up unit. When the pulse enters the control unit, it signals the control unit to interrupt the ignition primary circuit. This causes the primary circuit to collapse and begins the induction of the magnetic lines of force from the primary side of the coil into the secondary side of the coil. This induction provides the required voltage to fire the spark plugs.

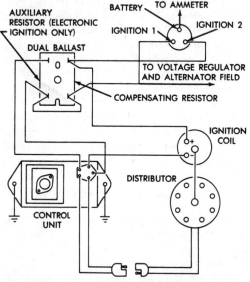

Schematic of the 1972–76 electronic ignition system except 1976 models with Lean Burn System

The advantages of this system are that the transistors in the control unit can make and break the primary ignition circuit much faster than conventional ignition points can, and higher primary voltage can be utilized, since this system can be made to handle higher voltage without adverse effects, whereas ignition breaker points cannot. The quicker switching time of this system allows longer coil primary circuit saturation time and longer induction time when the primary circuit collapses. This increased time allows the primary circuit to build up more current and the secondary circuit to discharge more current.

INSPECTION

NOTE: *If you own a 1972 Plymouth, and you aren't sure if it has electronic ignition or not, look for a double primary lead from the distributor, a dual ballast resistor, and a control unit located either on the left wheel arch or the firewall. If your car doesn't have these, it has a conventional ignition system. Cars made from 1973-on use electronic ignition exclusively.*

1. Inspect all of the secondary cables which run to the coil, distributor, and spark plugs, for signs of cracks or loose connections. Replace any cables necessary.

2. Check the wiring harness and connections at the control unit for tightness.

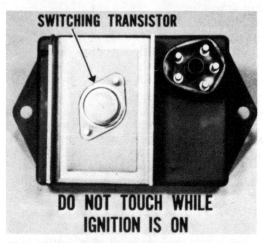

The switching transistor can give you a healthy shock if it is touched with the engine running

CAUTION: *Do not touch the large switching transistor on the outside of the control unit when the ignition is on; an electrical shock will result.*

3. Inspect the distributor cap and rotor for cracks. Inspect the cap terminals and rotor tip for excessive burning or corrosion; replace if required.

AIR GAP (RELUCTOR-TO-POLE PIECE)

The air gap is preset at the factory and because there is no physical contact between the reluctor teeth and the pole piece, it should not require adjustment. However, if the car won't start properly (or at all), check the air gap with a *nonmagnetic* feeler gauge, as outlined in the electronic ignition troubleshooting chart. Do this only as a last resort if no other trouble can be found, or if the components inside of the distributor have been disturbed. As a rule, the less tampering that you do with the electronic ignition

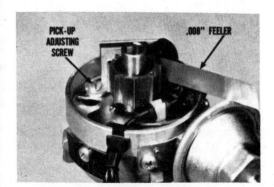

PICK-UP ADJUSTING SCREW

.008" FEELER

Use a piece of brass shim to measure the reluctor air gap

system, the less likely you are to have trouble with it.

If you really get stuck and can't find anything wrong with the electronic ignition, but you suspect that it is the cause of the trouble, check with your local dealer, he has a special test instrument which will allow the system to be checked and trouble in it traced.

DWELL ANGLE

The amount of dwell is determined by the circuit inside the electronic ignition control unit. There is no need to measure dwell, since it cannot be adjusted. Checking for proper operation of the circuitry which determine dwell requires the use of a special test instrument. If such a problem is suspected, check with your dealer's service department as he will have the necessary instrument to perform the tests.

Ignition Timing

Ignition timing is the measurement in degrees of crankshaft rotation of the instant the spark plugs in the cylinders fire, in relation to the location of the piston, while the piston is on its compression stroke.

Ignition timing is adjusted by loosening the distributor locking device and turning the distributor in the engine.

Ideally, the air/fuel mixture in the cylinder will be ignited (by the spark plug) and just beginning its rapid expansion as the piston passes top dead center (TDC) of the compression stroke. If this happens, the piston will be beginning the power stroke just as the compressed (by the movement of the piston) and ignited

(by the spark plug) air/fuel mixture starts to expand. The expansion of the air/fuel mixture will then force the piston down on the power stroke and turn the crankshaft.

It takes a fraction of a second for the spark from the plug to completely ignite the mixture in the cylinder. Because of this, the spark plug must fire before the piston reaches TDC, if the mixture is to be completely ignited as the piston passes TDC. This measurement is given in degrees (of crankshaft rotation) *before* the piston reaches *top dead center* (BTDC). If the ignition timing setting for your engine is six degrees (6°) BTDC, this means that the spark plug must fire at a time when the piston for that cylinder is 6° before top dead center of its compression stroke. However, this only holds true while your engine is at idle speed.

As you accelerate from idle, the speed of your engine (rpm) increases. The increase in rpm means that the pistons are now traveling up and down much faster. Because of this, the spark plugs will have to fire even sooner if the mixture is to be completely ignited as the piston passes TDC. To accomplish this, the distributor incorporates means to advance the timing of the spark as engine speed increases.

The distributor in your Plymouth has two means of advancing the ignition timing. One is called centrifugal advance and is actuated by weights in the distributor. The other is called vacuum advance and is controlled by that large circular housing on the side of the distributor.

Because these devices change ignition timing, it is necessary to disconnect and plug the vacuum lines from the distributor when setting the basic ignition timing.

If ignition timing is set too far advanced (BTDC), the ignition and expansion of the air/fuel mixture in the cylinder will try to force the piston down the cylinder while it is still traveling upward. This causes engine "ping," a sound which resembles marbles being dropped into an empty tin can. If the ignition timing is too far retarded (after, or ATDC), the piston will have already started down on the power stroke when the air/fuel mixture ignites and expands.

This will cause the piston to be forced down only a portion of its travel. This will result in poor engine performance and lack of power.

Ignition timing adjustment is checked with a timing light. This instrument is connected to the number one (no. 1) spark plug of the engine. The timing light flashes every time an electrical current is sent from the distributor, through the no. 1 spark plug wire, to the spark plug. The crankshaft pulley and the front cover of the engine are marked with a timing pointer and a timing scale. When the timing pointer is aligned with the "0" mark on the timing scale, the piston in no. 1 cylinder is at TDC of its compression stroke. With the engine running, and the timing light aimed at the timing pointer and timing scale, the stroboscopic flashes from the timing light will allow you to check the ignition timing setting of the engine. The timing light flashes every time the spark plug in the no. 1 cylinder of the engine fires. Since the flash from the timing light makes the crankshaft pulley seem stationary for a moment, you will be able to read the exact position of the piston in the no. 1 cylinder on the timing scale on the front of the engine.

If you own a Plymouth which has electronic ignition, you will very rarely, if ever, have to adjust the timing. Dwell and component wear effect the timing of a regular distributor, since dwell is controlled electronically and none of the parts in the electronic distributor make physical contact, periodic timing adjustments are unnecessary. The only time that timing should be checked or adjusted, is if the distributor has been removed from the engine or any of its internal parts have been replaced.

ADJUSTMENT

1. Locate the marks on the engine pulley and front cover.

2. Clean the timing scale and pulley mark. Mark the proper timing scale degree mark (See the "Tune-Up Specifications" chart) and the pulley mark with white chalk or paint.

3. Attach a timing light in accordance with the manufacturer's instructions.

a. If the timing light has three wires, one (usually blue or green) is installed with an adapter between the end of the No. 1 spark plug lead and the spark plug.

CAUTION: *Do not use probes to puncture the spark plug leads, boots, or nipples; they will become permanently damaged.*

b. Connect the two other leads to the positive (+) battery terminal (red lead) and a suitable ground (black lead).

4. Allow the engine to reach normal operating temperature. Set the engine idle (see below for procedure) to the figure given in the "Tune-Up Specifications" chart.

CAUTION: *Make sure that the timing light wires are clear of the fan, before starting the engine.*

5. Disconnect the vacuum line *at* the distributor and plug it with the pointed end of pencil or a golf tee.

6. Aim the timing light at the timing marks on the front of the engine. If the paint marks which you put on the degree scale and the pulley mark align when the timing light flashes, the timing is correct. Disconnect the timing light.

7. If the marks do not align, loosen

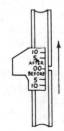

Timing marks—slant six

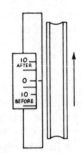

Timing marks—V8

the distributor clamp screw just enough to allow the distributor housing to be rotated, and then do one of the following:

a. If the flash occurs when the timing mark on the pulley is located *be-fore* (advanced) the correct degree mark on the scale (in the direction of engine rotation), turn the distributor housing in the direction of rotor rotation.

b. If the flash occurs when the mark on the pulley is located *after* (retarded) the correct degree mark on the scale (in the direction of engine rotation), turn the distributor housing in the opposite direction from which the rotor rotates.

CAUTION: *Do not use the vacuum advance unit to turn the distributor; it is not a handle.*

8. Once the proper timing specification has been obtained, tighten the distributor clamp screw, stop the engine, and disconnect the timing light.

9. Recheck and adjust the curb idle speed if necessary (see below), but do not readjust the timing after the curb idle speed has been changed.

Valve Lash Adjustment

6 CYLINDER

All six-cylinder engines use mechanical valve lifters (tappets). The valve lifter clearance (lash) should be adjusted at each tune-up.

To adjust the valve clearance, proceed in the following order:

1. Run the engine and allow it to reach normal operating temperature.

2. Remove the valve (rocker) cover, after disconnecting any of the emission control system hoses from it.

3. Place the No. 1 cylinder at top dead center (TDC) of its compression stroke, by turning the engine to the position at which the distributor rotor is pointing to the No. 1 cylinder terminal on the distributor cap and the mark on the crankshaft pulley aligns with "0" (TDC) on the timing scale. Both valves on the No. 1 cylinder should be fully *closed*, if they are not, turn the engine slightly in either direction until they are. Use either a remote starter switch to turn the engine or turn it with a wrench applied to the crankshaft pulley bolt. Be careful not to loosen the bolt.

4. Adjust both valves; it is important to note that the intake valves have intake

manifold passages running to them, while exhaust valves have exhaust manifold passages running to them, since the specified clearance is different for each. To adjust the valves proceed as follows:

a. Insert the proper size feeler gauge between the valve stem and the rocker arm. A slight drag should be felt, but the gauge should not buckle.

b. If the clearance is too large or too small, turn the adjusting nut on the rocker arm until the proper clearance is obtained.

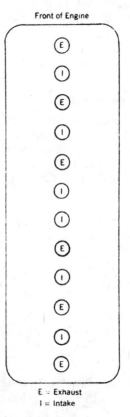

Front of Engine

E = Exhaust
I = Intake

Valve location—slant six

5. Turn the engine 120° clockwise to bring the next cylinder into position in the firing order sequence. The sequence is 1-5-3-6-2-4, so the next set of valves to be adjusted should be those for the No. 5 cylinder.

6. Repeat Steps 3–4 for the No. 5 cylinder, then go to the next cylinder in the firing order sequence (see Step 5). Keep repeating this procedure until all of the valves are adjusted.

7. When you are finished adjusting the valves, replace the valve cover and

connect any emission control system hoses which were removed.

V8 ENGINES

Non-adjustable hydraulic valve lifters are used on all other engines. These lifters are properly adjusted when the rocker arm shaft attaching bolts are torqued to specifications.

If the engine will not start after the rocker arm assembly has been replaced, spin the engine with the starter to allow the lifters to adjust to their proper height.

Carburetor Adjustments

The following are basic carburetor adjustments which are performed as part of the engine tune-up procedure, for more complete adjustment procedures, see Chapter 4, "Emission Controls and Fuel System."

IDLE SPEED/MIXTURE

To adjust the idle speed and mixture, particularly on later models, it is best to use an exhaust gas analyzer. This will insure that the proper level of emissions is maintained. However, if you do not have an exhaust gas analyzer, use the following procedure and eliminate those steps which pertain to the exhaust gas analyzer. When you have adjusted the carbu-

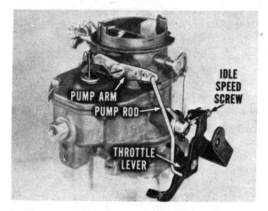

Curb idle speed adjustment without idle stop solenoid

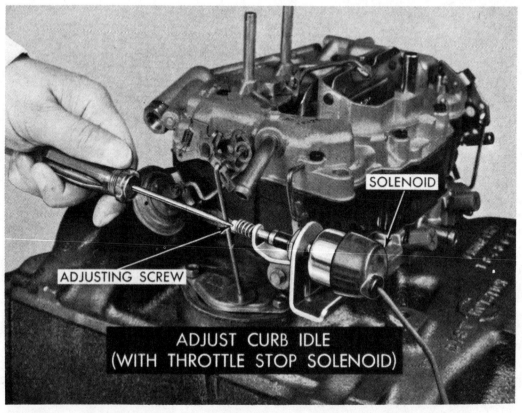

Curb idle speed adjustment with idle stop solenoid (antidieseling solenoid)

Carburetor with two idle mixture screws

Carburetor with single idle mixture screw

retor it would be wise to have it checked with an exhaust gas analyzer.

1. Leave the air cleaner installed.

2. Run the engine at fast idle speed to stabilize the engine temperature.

3. Make sure the choke plate is fully released.

4. Connect a tachometer to the engine, following the manufacturer's instructions.

5. Connect an exhaust gas analyzer and insert the probe as far into the tailpipe as possible. On vehicles with dual exhaust, insert the probe into the left side pipe, since this is the side with the heat riser.

6. Check the ignition timing and set it to specification if necessary.

7. If equipped with air conditioning, turn the air conditioner off. On 6-cylinder engines, turn the high beam lights on.

8. Put the transmission in Neutral. Make sure the hot idle compensator valve is fully seated.

9. If equipped with a distributor vacuum control valve, place a clamp on the line between the valve and the intake manifold.

10. Turn the engine idle speed adjusting screw in or out to adjust the idle speed to specification. If the carburetor is equipped with an electric solenoid throttle positioner, turn the solenoid adjusting screw in or out to obtain the specified rpm.

CAUTION: *On 1975–76 engines equipped with catalytic converters, be careful not to adjust the idle speed with the catalyst protection system solenoid; a dangerous engine overspeed condition could result. See Chapter 4 for a description of this system.*

11. Adjust the curb idle speed screw until it just touches the stop on the carburetor body. Back the curb idle speed screw out 1 full turn.

12. Turn each idle mixture adjustment screw 1/16 turn richer (counterclockwise). Wait 10 seconds and observe the reading on the exhaust gas analyzer. Continue this procedure until the meter indicates a definite increase in the richness of the mixture.

NOTE: *This step is very important when using an exhaust gas analyzer. A carburetor that is set too lean will cause a false reading from the analyzer, indicating a rich mixture. Because of this, the carburetor must first be known to have a rich mixture to verify the reading on the analyzer.*

13. After verifying the reading on the meter, adjust the mixture screws to obtain an air-fuel ratio of 14.2. Turn the mixture screws clockwise (leaner) to raise the meter reading or counterclockwise (richer) to lower the meter reading.

14. If the idle speed changes as the mixture screws are adjusted, adjust the speed to specification (see Step 10) and readjust the mixture so that the specified air/fuel ratio is maintained at the specified idle speed.

If the idle is rough, the screws may be adjusted independently provided that the 14.2 air/fuel ratio is maintained.

15. Remove the analyzer, the tachometer, and the clamp on the vacuum line.

ROUGH IDLE AND LOW SPEED SURGE

Rough idle and low-speed surge can be the result of improper balance of the idle mixture adjustment in the right and left carburetor bores. To correct this condition, perform the following operation.

1. Remove the plastic limiter caps from the idle mixture screws. Perform Steps 1–10 of the idle speed and mixture adjustment procedure.

2. Turn both idle mixture adjustment screws clockwise until they are lightly seated. On some models, the idle mixture screws have a prevailing torque feature which causes the screws to become more difficult to turn as they approach the seated position.

3. On Ball & Ball carburetors, turn both idle mixture screws 1½ turn coun- terclockwise. On Carter carburetors, turn both idle mixture screws 2–3 turns clock- wise.

4. Start the engine and perform Steps 11–15 of the idle speed and mixture ad- justment procedure.

NOTE: *In order to obtain a smooth idle, it is important that both mixture adjustment screws be adjusted an equal number of turns from the fully seated position.*

5. Install the caps on the idle mixture screws.

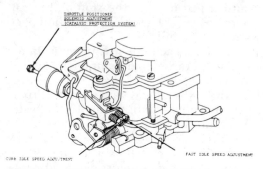

1975 Carter BBD Catalyst Protection System solenoid

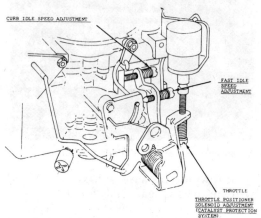

1975 Holley 2245 Catalyst Protection System solenoid

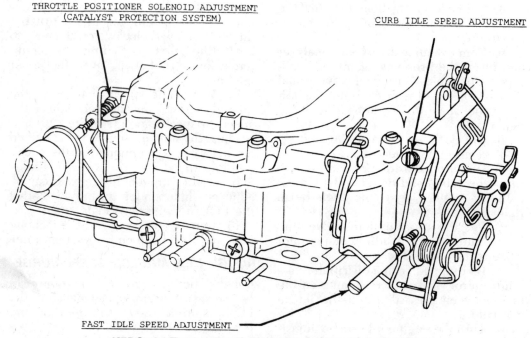

1975 Carter Thermo-Quad® Catalyst Protection System solenoid

Engine and Engine Rebuilding

ENGINE ELECTRICAL

Distributor

REMOVAL AND INSTALLATION

All Engines

1. Disconnect the vacuum advance line at the distributor.

2. Disconnect the primary wire at the coil. On electronic ignitions, disconnect the pick-up lead wire at the wiring harness connector.

3. Unfasten the distributor cap retaining clips and lift them off the cap.

4. Mark the distributor body and the engine block to indicate the position of the distributor in the block. Also, scribe a mark on the edge of the distributor housing to indicate the position of the rotor on the distributor. These marks can be used as guides when installing the distributor in a correctly timed engine.

5. Remove the distributor hold-down clamp screw and clamp.

6. Carefully lift the distributor out of the block. The shaft will rotate slightly as the distributor gear disengages.

If the crankshaft has not been rotated while the distributor was removed from the engine, installation is the reverse of the removal procedure. Use the reference marks made before removal to correctly position the distributor in the block. The shaft may have to be rotated slightly to engage the cam gear (6-cyl.) or intermediate shaft gear (V-8). Check the point gap and, before connecting the vacuum advance line, adjust the ignition timing (see "Tune-Up Procedures").

If the crankshaft was rotated or otherwise disturbed (e.g., during engine rebuilding) after the distributor was removed, proceed as follows.

INSTALLATION—ENGINE DISTURBED

Slant-Six

1. Remove No. 1 spark plug and, with the thumb closing the hole, rotate the engine until No. 1 piston is up on compression at top dead center. This is determined by the pressure on the thumb and the "0" mark on the crankshaft pully hub being aligned with the timing pointer.

2. Rotate the rotor to a position just ahead of the No.1 distributor cap terminal.

3. Lower the distributor into the opening, engaging distributor gear with

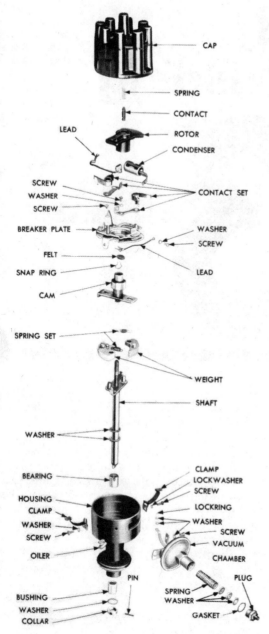

Conventional breaker point distributor—exploded view

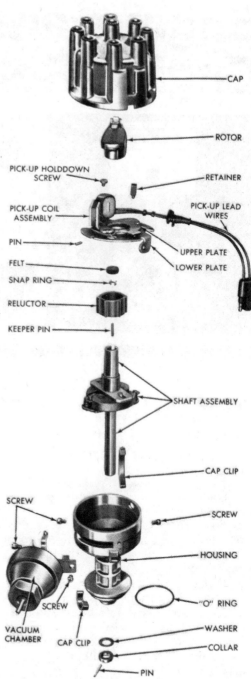

Electronic ignition distributor—exploded view

drive gear on camshaft. With distributor fully seated on engine, rotor should be under the cap No. 1 tower with distributor contact points just opening.

4. Install cap, tighten hold-down arm screw and check timing with a timing light, and dwell with a dwell meter.

V8 Engines

1. Rotate the crankshaft until No. 1 cylinder is at top dead center (TDC) of the compression stroke. This can be de-

termined by holding your finger over the No. 1 spark plug hole until compression is felt and both valves are closed. The pointer on the chain case cover should be over the "0" mark on the crankshaft pulley. The slot in the intermediate shaft which carries the gear that drives the oil pump and the distributor, should be parallel with the crankshaft.

2. Hold the distributor over the mounting pad on the cylinder block so that the distributor body flange coincides with the mounting pad and the rotor points to the No. 1 cylinder firing position.

3. Install the distributor while holding the rotor in position, allowing it to move only enough to engage the slot in the drive gear.

Firing Order

To avoid confusion, replace spark plug wires one at a time.

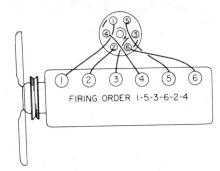

FIRING ORDER 1-5-3-6-2-4

Slant six

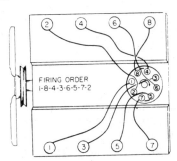

FIRING ORDER 1-8-4-3-6-5-7-2

318 and 360 cu in. V8s

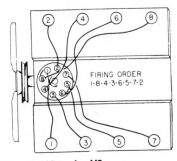

FIRING ORDER 1-8-4-3-6-5-7-2

383, 400, and 440 cu in. V8s

Alternator

ALTERNATOR PRECAUTIONS

Some precautions that should be taken into consideration when working on this,

or any other, AC charging system are as follows:

1. Never reverse battery polarity.

2. When installing a battery, always connect the grounded terminal first.

3. Never disconnect the battery while the engine is running.

4. If the molded connector is disconnected from the alternator, do not ground the hot wire.

5. Never run the alternator with the main output cable disconnected.

6. Never electric arc-weld around the car without disconnecting the alternator.

7. Never apply any voltage in excess of battery voltage during testing.

8. Never "jump" a battery for starting purposes with more than 12 volts.

REMOVAL AND INSTALLATION

1. Disconnect the battery ground cable at the battery negative terminal.

2. Disconnect and tag the alternator output (BATT) and field (FLD) leads and disconnect the ground wire.

3. Loosen the alternator adjusting bolt and swing the alternator in toward the engine. Disengage the alternator drive belt.

4. Remove the alternator mounting bolts and remove the alternator from the vehicle.

5. Installation is the reverse of removal. Be sure to connect all ground wires and leads securely.

6. Adjust the belt tension.

BELT TENSION ADJUSTMENT

The belt tension adjustment is correct when there is ½ in. deflection under thumb pressure in the middle of the longest span. Loosen the alternator mounting bolt and pull the alternator by hand, testing the deflection at the same time. When approximately ½ in. of deflection is reached, tighten the alternator mounting bolt while holding the alternator in place.

NOTE: *Do not put too much tension on the alternator or the belt, as this will tend to wear the bearings in the alternator.*

Voltage Regulator

The function of the voltage regulator is to limit the output voltage by controlling the flow of current in the alternator rotor field coil which, in effect, controls the strength of the rotor magnetic field. On all models, the output voltage is limited by two different types of voltage regulators.

The 1968–69 models use a mechanical voltage regulator, i.e., the regulator has contact points which are adjustable.

All 1970–76 models are equipped with a solid-state (silicon transistor) voltage regulator which is not adjustable.

REMOVAL AND INSTALLATION

Both types of voltage regulators can be removed and installed by using the same procedure.
1. Disconnect the cables from the battery posts.
2. Label and disconnect each electrical lead from the voltage regulator.
3. Remove the regulator by unfastening its securing screws.
4. Installation is the reverse of the above. Be sure that the electrical leads are connected to the correct terminals and that all connections are clean and tight. If possible, test the voltage regulator to be sure that the output is correct.

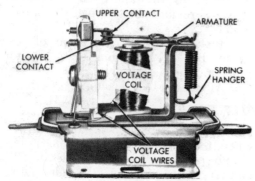

Mechanical voltage regulator with the cover removed

VOLTAGE REGULATOR ADJUSTMENTS

1968–69

Adjust the *upper contact* voltage setting as follows:
1. Remove the regulator cover.
2. Use an insulated tool to adjust the upper contact voltage as necessary by bending the regulator lower spring hanger *down* to *increase* the voltage setting or *up to decrease* the voltage setting.

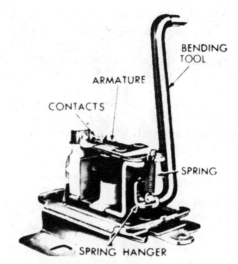

Spring tension adjustment

3. If the voltage reading is now correct, refit the regulator cover.
Adjust the *lower contact* voltage setting as follows:
1. Remove the regulator cover.
2. Measure the lower contact point gap with a feeler gauge. The lower contact gap should be 0.014 in. plus or minus 0.002 in. If necessary, adjust the contact gap by bending the lower stationary contact bracket while making sure that the contacts remain in alignment.
3. If the voltage reading is now correct, refit the regulator cover. If the lower contact gap is correct but the voltage reading is still outside the 0.2–0.7 volt increase, continue this procedure to adjust the lower contacts air gap.
4. Connect a small dry cell and test lamp in series with the IGN and FLD terminals of the voltage regulator.
5. Insert a 0.048 in. wire gauge between the regulator armature and the core of the voltage coil next to the stop pin on the armature.
6. Press down on the armature (not on the contact reed) until the armature contacts the wire gauge. The upper contacts should just open and the test lamp should be dim.
7. Insert a 0.052 in. wire gauge between the armature and the voltage coil core, next to the stop pin on the armature.
8. Press down on the armature until it contacts the wire gauge. The upper

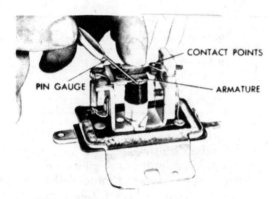

Measuring air gap

contacts should remain closed and the test lamp should remain bright.

9. To obtain the correct difference of 0.2–0.7 volt increase of the lower contact's voltage over the upper contact voltage, adjust the lower contact's air gap by loosening the stationary contact bracket screw and moving the bracket up or down to obtain the proper air gap setting as follows:

a. If the difference is above 0.7 volt (V), reduce the air gap to a minimum of 0.045 in. with the contacts open and the test lamp dim. At 0.048 in., the contacts should close and the test lamp should be bright.

b. If the difference is below 0.2 V, increase the air gap to a maximum of 0.055 in., the contacts should be open and the test lamp should be dim. NOTE: *Be certain that the air gap is measured with the stationary contact bracket attaching screw fully tightened.*

10. When all adjustments are complete, refit the regulator cover.

TRANSISTORIZED VOLTAGE REGULATOR

This voltage regulator maintains correct charging voltage by varying the duty cycle of a series of pulses to the alternator field. The pulse frequency is controlled by the ignition frequency of the engine. The regulator has no moving parts and requires no adjustment after it is set internally at the factory. If the unit is found to be defective, it must be removed and replaced with a new regulator.

Voltage Regulator Test 1970–76

1. Clean the battery terminals and check the specific gravity of the battery electrolyte. If the specific gravity is below 1.200, charge the battery before performing the voltage regulator test as

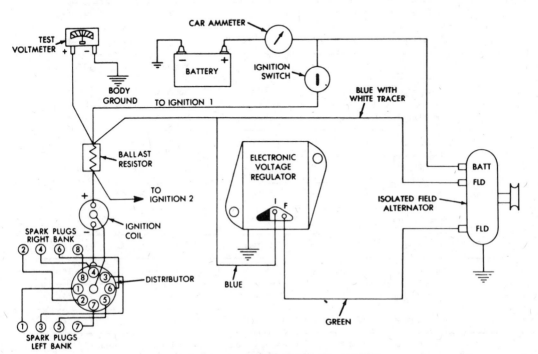

Transistorized voltage regulator test circuit

it must be above 1.200 to allow a prompt, regulated voltage check.

2. Connect the positive lead of the test voltmeter to the ignition no. 1 terminal of the ballast resistor. (The end with one or two blue wires connected to it.)

3. Connect the voltmeter negative lead to a good body ground.

4. Start and operate the engine at 1,250 rpm with all lights and accessories switched off. Observe the voltmeter reading. The regulator is working properly if the voltage readings are in accordance with the following chart.

Ambient Temp Near Regulator	Voltage Range
Below 20° F	14.3–15.3
80° F	13.8–14.4
140° F	13.3–14.0
Above 140° F	Less than 13.8

5. a. If the voltage reading is below the specified limits, check for a bad voltage regulator ground. Using the low voltage scale of the test meter, check for a voltage drop between the regulator cover and body.

Alternator and Regulator Specifications

		ALTERNATOR			REGULATOR						
						Field Relay			Regulator		
Year	Usage	Field Current @ 12 V	Output (amps)	Part No. or Manufacturer	Air Gap (in.)	Point Gap (in.)	Volts to Close	Air Gap (in.)	Point Gap (in.)	Volts @ 75°
'68–'69	6 Cyl Models	2.38–2.75	26 ± 3	2098300①	.050①	.014	13.8	.015	.050	13.8–14.4
	V8 Std.—All	2.38–2.75	34.5 ± 3	2098300①	.050①	.014	13.8	.015	.050	13.8–14.4
	Heavy Duty, A/C	2.38–2.75	44 ± 3②	2098300①	.050①	.014	13.8	.015	.050	13.8–14.4
'70–'71	6 Cyl Models	2.38–2.75	26 ± 3	3438150	Not Adjustable					13.8–14.4
	V8 Std.—All	2.38–2.75	34.5 ± 3	3438150	Not Adjustable					13.8–14.4
	Heavy Duty, A/C	2.38–2.75	44.5 ± 3	3438150	Not Adjustable					13.8–14.4
	Special Equip.	2.38–2.75	51 ± 3	3438150	Not Adjustable					13.8–14.4
'72–'73	V8 Std.	2.5–3.1	39	3438150	Not Adjustable					13.8–14.4
	Heavy Duty, A/C	2.5–3.1	50	3438150	Not Adjustable					13.8–14.4
	Special Equip.	2.5–3.1	60	3438150	Not Adjustable					13.8–14.4
'74	V8 Std.	2.5–3.1	50	3438150	Not Adjustable					13.8–14.4
	Special Equip.	2.5–3.1	65	3438150	Not Adjustable					13.8–14.4
'75	V8 Std.	2.5–3.7	50	3755960	Not Adjustable					14.0–14.6
	Special Equip.	2.5–3.7	65	3755960	Not Adjustable					14.0–14.6
	w/Rear defogger	4.75–6.0	100	3755960	Not Adjustable					14.0–14.6
'76	V8 Std.	4.5–6.5	60	3874510	Not Adjustable					14.0–14.6
	Special Equip.	4.5–6.5	65	3874510	Not Adjustable					14.0–14.6
	w/Rear Defogger	4.75–6.0	100	3874510	Not Adjustable					14.0–14.6

① Chrysler built used interchangeably with #2444900, which is Essex Wire built. Air gap setting is .032–.042 in., all other dimensions are identical with #2098300
② 51 amp special equipment model available

b. If the reading is still low, switch off the ignition and disconnect the voltage regulator connector. Switch on the ignition but do not start the car. Check for battery voltage at the wiring harness terminal connected to the blue and green leads. *Disconnect the wiring harness from the voltage regulator when checking the leads.* Switch off the ignition.

c. If there is no voltage at either lead, the problem is in the vehicle wiring or alternator field circuit. If voltage is present, change the voltage regulator and repeat Step 4.

6. If the voltage reading is above the specified limits, check the ground between the voltage regulator and the vehicle body, and between the vehicle body and the engine. Check the ignition switch circuit between the switch battery terminal and the voltage regulator. If the voltage reading is still high (more than ½ V above the specified limits), change the voltage regulator and repeat Step 4.

7. Remove the test voltmeter.

Starter

Chrysler Corporation cars use two types of starters: a direct-drive type or a reduction gear type. The reduction gear type may be identified by the battery terminal on the starter being installed at a 45° angle in relation to the case; the direct-drive type starter battery terminal is parallel to the starter case.

Both types have solenoids which are mounted directly on the starter assembly. The starter must be removed from the car to service the solenoid and motor brushes.

REMOVAL AND INSTALLATION

1. Disconnect the ground cable at the battery.

2. Remove the cable from the starter.

3. Disconnect the solenoid leads at their solenoid terminals.

4. Remove the starter securing bolts and withdraw the starter from the engine flywheel housing. On some models with automatic transmissions, the oil cooler

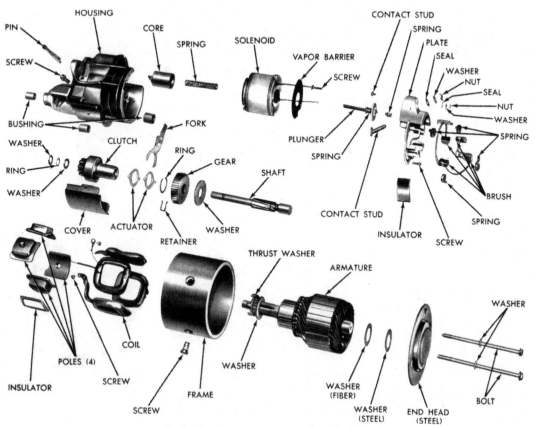

Reduction gear starter—exploded view

tube bracket will interfere with starter removal. In this case, remove the starter securing bolts, slide the cooler tube bracket off the stud, then withdraw the starter.

5. Installation is the reverse of the above. Be sure that the starter and flywheel housing mating surfaces are free of dirt and oil.

SOLENOID AND BRUSH SERVICE

Reduction Gear Starter

1. Remove the starter from the car and support the starter gear housing in a vise with soft jaws. DO NOT CLAMP.

2. Remove the two thru-bolts and the starter end assembly.

3. Carefully pull the armature up and out of the gear housing and the starter frame and field assembly.

4. Carefully pull the frame and field assembly up just enough to expose the terminal screw (which connects the series field coils to one pair of motor brushes) and support it with two blocks.

5. Support the terminal by placing a finger behind the terminal and remove the terminal screw.

6. Unwrap the shunt field coil lead from the other starter brush terminal. Unwrap the solenoid lead wire from the brush terminals.

7. Remove the steel and fiber thrust washer.

8. Remove the nut, steel washer, and insulating washer from the solenoid terminal.

9. Straighten the solenoid wire and remove the brush holder plate with the brushes and solenoid as an assembly.

10. Inspect the starter brushes. Brushes that are worn more than one-half the length of new brushes or are oil-soaked, should be replaced.

11. Assemble the starter using the reverse of the above procedure. When re-soldering the shunt field and solenoid leads, make a strong, low-resistance connection using a high-temperature solder and resin flux. *Do not break the shunt field wire units when removing and installing the brushes.*

Direct-Drive Starter

1. Remove the starter from the car and support it in a vise with soft jaws. DO NOT CLAMP.

2. Remove the thru-bolts and tap the commutator and head from the field frame.

3. Remove the thrust washers from the armature shaft.

4. Lift the brush holder springs and remove the brushes from the brush holders. Remove the brush plate.

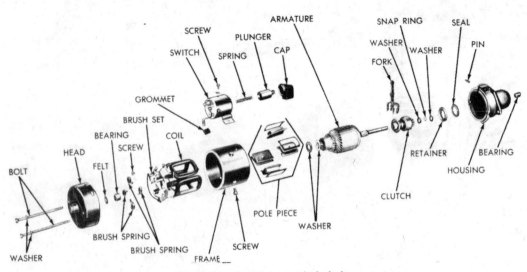

Direct drive starter—exploded view

5. Disconnect the field coil leads at the solenoid connector.

6. Inspect the starter brushes. Brushes that are worn more than one-half the length of new brushes or are oil-soaked should be replaced. To replace the brushes, continue this procedure as follows.

7. Remove the ground brush terminal screw and carefully remove the ground brush set to prevent breaking the shunt field lead. Remove the shunt field lead from the old brush set to provide as much length as possible.

8. Remove the field terminal plastic covering and remove the old brushes. Use side cutters to break the weld by rolling the stranded wire off the terminal.

9. Drill a 0.174–0.184 in. hole in the series coil terminal $^3/_{16}$ in. from the top of the terminal to the centerline of the hole. (Use a no. 16 drill.)

CAUTION: *Do not damage the field coil during the drilling operation.*

10. Attach the insulated brush set to the series field terminal with a flat washer and no. 8 self-tapping screw. Attach the shunt field lead to the new ground brush set by making a loop around the terminal and soldering the lead to the terminal with resin core solder.

11. Attach the ground brush terminal to the field frame with the securing screw. Fold the extra shunt field lead back along the brush lead and secure it with electrical tape.

12. Assemble the starter using the reverse of Steps 1–5.

Battery

REMOVAL AND INSTALLATION

1. Protect the paint finish with fender covers.

2. Disconnect the cable from the negative (–) battery post *first;* then disconnect the cable from the positive (+) post.

3. Remove the battery hold-down clamp and remove the battery from the vehicle. If the car has a battery cover (1975–76 only), lift the cover off the battery.

4. Inspect the battery carrier and the fender panels for battery acid damage.

5. If the battery is to be reinstalled, scrub the top of the battery with a solution of baking soda and warm water, using a wire brush to remove any heavy deposits. Rinse with clean, warm water.

CAUTION: *Keep the cleaning solution and water out of the battery cells.*

6. Examine the battery case and cover for cracks.

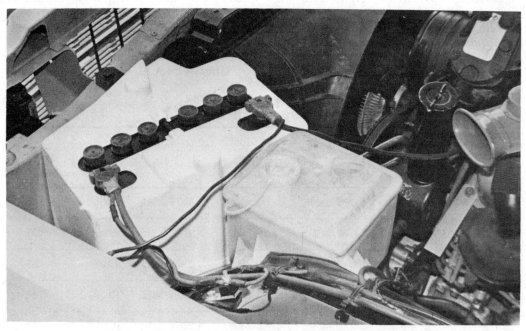

Battery with protective heat shield; used on most 1975–76 models

Battery and Starter Specifications

| Year | Engine Displacement (cu in.) | BATTERY | | | STARTER | | | | | | | Brush Spring Tension (oz) |
| | | Ampere Hour Capacity | Volts | Terminal Grounded | Lock Test | | | No-Load Test | | | | |
					Amps	Volts	Torque (ft lbs)	Amps	Volts	RPM		
'68–'69	6-225, 8-318	48	12	Neg.	400–450	4	——	90	11	1,925–2,600		32–36
	8-383, 440	70	12	Neg.	400–450	4	——	90	11	1,925–2,600		32–36
'70–'74	6-225, 8-318	46	12	Neg.	400–450	4	——	90	11	1,925–2,600		32–36
	8-360, 383, 400	59②	12	Neg.	400–450	4	——	90	11	1,925–2,600		32–36
	8-440	70	12	Neg.	400–450	4	——	90	11	1,925–2,600		32–36
'75–'76	8-318	48①	12	Neg.	475–550	4	——	90	11	3,700		32–36
	8-360, 400	70①	12	Neg.	475–550	4	——	90	11	5,700		32–36
	8-440	85	12	Neg.	475–550	4	——	90	11	5,700		32–36

① 85 amp-hour battery optional ② 55 amps for 1972 8-360, 400

7. Clean the battery posts and cable terminals with a wire brush. Replace damaged or worn cables.

8. Install the battery in the car, after replacing the protective cover (1975–76 only). Tighten the hold-down clamp nuts.

9. Connect the cables in the *exact* reverse order of removal. Be sure that the right cable goes to the right terminal; reversing the cables will result in damage to the car's electrical system. Coat all connections with petroleum jelly or silicone spray to prevent corrosion.

10. If the electrolyte level is low, fill the battery to the recommended level with distilled water (see Chapter 1).

ENGINE MECHANICAL

Design

The standard equipment engine in most Plymouth models until 1972 is the 225 cu in. slant-six. Although this engine has a very long stroke by modern standards, it presents a low profile because the block is canted 30° to the right.

The 318 and 360 cu in. engines make up Chrysler's A block series of V8s. The

General Engine Specifications

Year	Engine No. Cyl Displacement Cu In.	Carburetor Type	Advertised Horsepower @ rpm ∎	Advertised Torque @ rpm (ft lbs) ∎	Bore and Stroke (in.)	Advertised Compression Ratio	Oil Pressure @ 2050 rpm
'68	6-225	1 bbl	145 @ 4000	215 @ 2400	3.400 x 4.125	8.40 : 1	55
	8-318	2 bbl	230 @ 4400	340 @ 2400	3.910 x 3.310	9.20 : 1	55
	8-383	2 bbl	290 @ 4400	390 @ 2800	4.250 x 3.375	9.20 : 1	55
	8-383	4 bbl	330 @ 5000	425 @ 3200	4.250 x 3.375	10.00 : 1	55
	8-440	4 bbl	350 @ 4400	480 @ 2800	4.320 x 3.750	10.01 : 1	55

Year	Engine No. Cyl Displacement Cu In.	Carburetor Type	Advertised Horsepower @ rpm ■	Advertised Torque @ rpm (ft lbs) ■	Bore and Stroke (in.)	Advertised Compression Ratio	Oil Pressure @ 2050 rpm
	8-440 HP	4 bbl	375 @ 4600	480 @ 3200	4.320 x 3.750	10.01 : 1	55
'69	6-225	1 bbl	195 @ 4000	215 @ 2400	3.400 x 4.125	8.40 : 1	55
	8-318	2 bbl	230 @ 4400	340 @ 2400	3.910 x 3.310	9.20 : 1	55
	8-383	2 bbl	290 @ 4400	390 @ 2800	4.250 x 3.375	9.20 : 1	55
	8-383	4 bbl	330 @ 5000	425 @ 3200	4.250 x 3.375	10.00 : 1	55
	8-383	4 bbl	335 @ 5200	425 @ 3400	4.250 x 3.375	10.00 : 1	55
	8-440	4 bbl	350 @ 4400	480 @ 2800	4.320 x 3.750	10.01 : 1	55
	8-440 HP	4 bbl	375 @ 4600	480 @ 3200	4.320 x 3.750	10.01 : 1	55
'70	6-225	1 bbl	145 @ 4000	215 @ 2400	3.400 x 4.125	8.40 : 1	55
	8-318	2 bbl	230 @ 4400	320 @ 2000	3.910 x 3.310	8.80 : 1	55
	8-383	2 bbl	290 @ 4400	390 @ 2800	4.250 x 3.375	8.70 : 1	55
	8-383	4 bbl	330 @ 5000	425 @ 3200	4.250 x 3.375	9.50 : 1	55
	8-440	4 bbl	350 @ 4400	480 @ 2800	4.320 x 3.750	9.70 : 1	55
	8-440 HP	4 bbl	375 @ 4600	480 @ 3200	4.320 x 3.750	9.70 : 1	55
'71	6-225	1 bbl	145 @ 4000	215 @ 2400	3.400 x 4.125	8.40 : 1	55
	8-318	2 bbl	230 @ 4400	320 @ 2000	3.910 x 3.310	8.60 : 1	55
	8-360	2 bbl	255 @ 4400	360 @ 2400	4.000 x 3.580	8.70 : 1	55
	8-383	2 bbl	275 @ 4400	375 @ 2800	4.250 x 3.375	8.50 : 1	55
	8-383 HP	4 bbl	300 @ 4800	410 @ 3400	4.250 x 3.375	8.50 : 1	55
	8-440	4 bbl	335 @ 4400	460 @ 3200	4.320 x 3.750	8.50 : 1	55
	8-440 HP	4 bbl	370 @ 4600	480 @ 3200	4.320 x 3.750	9.50 : 1	55
'72	8-318	2 bbl	150 @ 4000	260 @ 1600	3.910 x 3.310	8.60 : 1	55
	8-360	2 bbl	175 @ 4000	285 @ 2400	4.000 x 3.580	8.80 : 1	55
	8-400	2 bbl	190 @ 4400①	310 @ 2400②	4.340 x 3.380	8.20 : 1	55
	8-400	4 bbl	255 @ 4800③	340 @ 3200④	4.340 x 3.380	8.20 : 1	55

General Engine Specifications (cont.)

Year	Engine No. Cyl Displacement Cu In.	Carburetor Type	Advertised Horsepower @ rpm ■	Advertised Torque @ rpm (ft lbs) ■	Bore and Stroke (in.)	Advertised Compression Ratio	Oil Pressure @ 2050 rpm
	8-440	4 bbl	225 @ 4400⑤	345 @ 3200⑥	4.320 x 3.750	8.20 : 1	55
'73	8-318	2 bbl	150 @ 3600	265 @ 2000	3.910 x 3.310	8.6 : 1	55
	8-360	2 bbl	170 @ 4000	285 @ 2400	4.000 x 3.580	8.4 : 1	55
	8-400	2 bbl	185 @ 3600	310 @ 2400	4.340 x 3.380	8.2 : 1	55
	8-400HP	4 bbl	260 @ 4800	335 @ 3600	4.340 x 3.380	8.2 : 1	55
	8-440	4 bbl	220 @ 3600	350 @ 2400	4.320 x 3.750	8.2 : 1	55
	8-440HP	4 bbl	275 @ 4800	380 @ 3200	4.320 x 3.750	8.2 : 1	55
'74	8-360	2 bbl	160 @ 3600	285 @ 2400	4.000 x 3.580	8.4 : 1	55
	8-400	2 bbl	175 @ 3600	305 @ 2400	4.340 x 3.380	8.2 : 1	55
	8-400	4 bbl	185 @ 3600	310 @ 2400	4.340 x 3.380	8.2 : 1	55
	8-440	4 bbl	220 @ 3600	350 @ 2400	4.320 x 3.750	8.2 : 1	55
	8-440HP	4 bbl	280 @ 4800	380 @ 3200	4.320 x 3.750	8.2 : 1	55
'75	8-318	2 bbl	150 @ 4000⑦	260 @ 1600⑧	3.910 x 3.310	8.5 : 1	55
	8-360	2 bbl	180 @ 4000	290 @ 2400	4.000 x 3.580	8.4 : 1	55
	8-360	4 bbl	190 @ 4000	270 @ 3200	4.000 x 3.580	8.4 : 1	55
	8-400	2 bbl	175 @ 4000	300 @ 2400	4.340 x 3.380	8.2 : 1	55
	8-400⑬	4 bbl	195 @ 4000⑨	285 @ 3200⑩	4.340 x 3.380	8.2 : 1	55
	8-440	4 bbl	215 @ 4000⑪	330 @ 3200⑫	4.320 x 3.750	8.2 : 1	55
'76	8-318	2 bbl	150 @ 4000	255 @ 1600	3.910 x 3.310	8.5 : 1	55
	8-360	2 bbl	170 @ 4000	280 @ 2400	4.000 x 3.580	8.4 : 1	55
	8-360⑬	4 bbl	175 @ 4000	270 @ 1600	4.000 x 3.580	8.4 : 1	55
	8-400	2 bbl	175 @ 4000	300 @ 2400	4.340 x 3.380	8.2 : 1	55
	8-400⑬	4 bbl	185 @ 3600	285 @ 3200	4.340 x 3.380	8.2 : 1	55
	8-400LB	4 bbl	210 @ 4400	305 @ 3200	4.340 x 3.380	8.2 : 1	55
	8-400HP	4 bbl	240 @ 4400	325 @ 3200	4.340 x 3.380	8.2 : 1	55
	8-440	4 bbl	205 @ 3600	320 @ 2000	4.320 x 3.750	8.2 : 1	55
	8-440HP	4 bbl	255 @ 4400	355 @ 3200	4.320 x 3.750	8.2 : 1	55

■ Beginning 1972, horsepower and torque are SAE net figures. They are measured at the rear of the transmission with all accessories installed and operating. Since the figures vary when a given engine is installed in different models, some are representative rather than exact.
① For California vehicles, advertised horsepower is 181 @ 4400 rpm
② For California vehicles, advertised torque is 305 @ 2400 rpm
③ For California vehicles, advertised horsepower is 246 @ 4800 rpm
④ For California vehicles, advertised torque is 335 @ 3200 rpm
⑤ For California vehicles, advertised horsepower is 216 @ 4400 rpm

⑥ For California vehicles, advertised torque is 340 @ 3200 rpm
⑦ For California vehicles, advertised horsepower is 135 @ 3600 rpm
⑧ For California vehicles, advertised torque is 245 @ 1600 rpm
⑨ For California vehicles, advertised horsepower is 185 @ 4000 rpm
⑩ For California vehicles, advertised torque is 285 @ 3200 rpm
⑪ For California vehicles, advertised horsepower is 210 @ 4000 rpm
⑫ For California vehicles, advertised torque is 320 @ 3200 rpm
HP High Performance or Police Interceptor
⑬ California only

Valve Specifications

Year	Engine No. Cyl Displacement (cu in.)	Seat Angle (deg)	Face Angle (deg)	Spring Test Pressure (lbs @ in.)	Spring Installed Height (in.)	STEM TO GUIDE Clearance (in.)		STEM Diameter (in.)	
						Intake	Exhaust	Intake	Exhaust
'68	6-225	45	45①	144 @ 1.31	1¹¹⁄₁₆	.0010–.0030	.0020–.0040	.3725	.3715
	8-318	45	45①	177 @ 1.31	1¹¹⁄₁₆	.0010–.0030	.0020–.0040	.3725	.3715
	8-383③	45	45	200 @ 1.44	1⅞	.0010–.0030	.0020–.0040	.3725	.3715
	8-383④	45	45	230 @ 1.41	1⅞	.0010–.0030	.0020–.0040	.3725	.3715
	8-440	45	45	200 @ 1.44	1⅞	.0010–.0030	.0020–.0040	.3725	.3715
	8-440	45	45	230 @ 1.41	1⅞	.0010–.0030	.0020–.0040	.3725	.3715
'69	6-225	45	45①	144 @ 1.31	1¹¹⁄₁₆	.0010–.0030	.0020–.0040	.3725	.3715
	8-318	45	45①	177 @ 1.31	1¹¹⁄₁₆	.0010–.0030	.0020–.0040	.3725	.3715
	8-383③	45	45	200 @ 1.44	1⅞	.0010–.0030	.0020–.0040	.3725	.3715
	8-383④	45	45	246 @ 1.36	1⅞	.0010–.0030	.0020–.0040	.3725	.3715
	8-440	45	45	200 @ 1.44	1⅞	.0010–.0030	.0020–.0040	.3725	.3715
	8-440⑤	45	45	246 @ 1.36	1⅞	.0010–.0030	.0020–.0040	.3725	.3715
'70	6-225	45	45①	144 @ 1.31	1¹¹⁄₁₆	.0010–.0030	.0020–.0040	.3725	.3715
	8-318	45	45①	177 @ 1.31	1¹¹⁄₁₆	.0010–.0030	.0020–.0040	.3725	.3715
	8-383③	45	45	200 @ 1.44	1⅞	.0010–.0030	.0020–.0040	.3727	.3717
	8-383④	45	45	246 @ 1.72	1⅞	.0015–.0032	.0025–.0042	.3722	.3712
	8-440	45	45	246 @ 1.72	1⅞	.0010–.0030	.0020–.0040	.3727	.3717
	8-440⑤	45	45	310 @ 1.38	1⅞	.0015–.0032	.0025–.0042	.3722	.3712
'71	6-225	45	45①	144 @ 1.31	1¹¹⁄₁₆	.0010–.0030	.0020–.0040	.3725	.3715
	8-318	45	45①	177 @ 1.31·	1¹¹⁄₁₆	.0010–.0030	.0020–.0040	.3725	.3715
	8-360	45	45①	177 @ 1.31	1¹¹⁄₁₆	.0010–.0030	.0020–.0040	.3725	.3715
	8-383③	45	45	200 @ 1.44	1⅞	.0010–.0030	.0020–.0040	.3727	.3717
	8-383④	45	45	246 @ 1.72	1⅞	.0015–.0032	.0025–.0042	.3722	.3712
	8-440	45	45	200 @ 1.44	1⅞	.0010–.0030	.0020–.0040	.3722	.3717
	8-440⑤	45	45	246 @ 1.72	1⅞	.0015–.0032	.0025–.0042	.3722	.3712

Valve Specifications (cont.)

Year	Engine No. Cyl Displacement (cu in.)	Seat Angle (deg)	Face Angle (deg)	Spring Test Pressure (lbs @ in.)	Spring Installed Height (in.)	STEM TO GUIDE Clearance (in.)		STEM Diameter (in.)	
						Intake	Exhaust	Intake	Exhaust
'72	8-318	45	45①	177 @ 1.31	1¹¹⁄₁₆	.0010–.0030	.0020–.0040	.3725	.3715
	8-360	45	45①	177 @ 1.31	1¹¹⁄₁₆	.0010–.0030	.0020–.0040	.3725	.3715
	8-400③	45	45	200 @ 1.44	1⅞	.0010–.0030	.0020–.0040	.3727	.3717
	8-400④	45	45	246 @ 1.72	1⅞	.0015–.0032	.0025–.0042	.3722	.3712
	8-440	45	45	200 @ 1.44	1⅞	.0010–.0030	.0020–.0040	.3727	.3717
	8-440⑤	45	45	246 @ 1.72	1⅞	.0015–.0032	.0025–.0042	.3722	.3712
'73	8-318	45	45②	189 @ 1.28	1²¹⁄₃₂	.0010–.0030	.0020–.0040	.3725	.3715
	8-360	45	45②	195 @ 1.24	1²¹⁄₃₂	.0010–.0030	.0020–.0040	.3725	.3715
	8-400③	45	45	200 @ 1.42	1⁵⁵⁄₆₄	.0010–.0027	⑦	.3727	⑨
	8-400④	45	45	234 @ 1.40	1⁵⁵⁄₆₄	.0015–.0032	⑧	.3722	⑩
	8-440	45	45	200 @ 1.42	1⁵⁵⁄₆₄	.0010–.0027	⑦	.3727	⑨
	8-440⑤	45	45	234 @ 1.40	1⁵⁵⁄₆₄	.0015–.0032	⑧	.3722	⑩
'74	8-360	45	45	238 @ 1.22⑭	1²¹⁄₃₂	.0010–.0030	.0025–.0045	.3725	.3710
	8-400	45	45	200 @ 1.43	1⁵⁵⁄₆₄	.0010–.0027	⑥	.3727	⑧
	8-400⑤	45	45	234 @ 1.40	1⁵⁵⁄₆₄	.0015–.0032	⑦	.3722	⑨
	8-440	45	45	234 @ 1.40	1⁵⁵⁄₆₄	.0015–.0032	⑦	.3722	⑨
'75–'76	8-318	⑪	⑫	185 @ 1.28⑭	1²¹⁄₃₂	.0010–.0030	.0020–.0040	.3725	.3715
	8-360③	⑪	⑫	195 @ 1.24	1²¹⁄₃₂	.0010–.0030	.0020–.0040	.3725	.3715
	8-360④	⑪	⑫	195 @ 1.24	1²¹⁄₃₂	.0015–.0035	.0020–.0040	.3720	.3715
	8-400	⑪	⑬	200 @ 1.43	1⁵⁵⁄₆₄	.0010–.0027	⑥	.3726	⑨
	8-440	⑪	⑬	200 @ 1.43	1⁵⁵⁄₆₄	.0010–.0027	⑥	.3726	⑨

① Exhaust 43°
② Exhaust 47°
③ 2 bbl carburetor
④ 4 bbl carburetor
⑤ Hi-Performance
⑥ Hot end—.0020–.0037, cold end—.0010–.0027
⑦ Hot end—.0025–.0042, cold end—.0015–.0032
⑧ Hot end—.3716, cold end—.3726
⑨ Hot end—.3711, cold end—.3721
⑩ 195 @ 1.24 on 2 bbl engine
⑪ Intake and exhaust, 44.5–45.0°
⑫ Intake 45.0–45.5°; exhaust 47.0–47.5°
⑬ Intake and exhaust 45.0–45.5°
⑭ 1976 318 exhaust valve spring—192 @ 1.25 in.

Crankshaft and Connecting Rod Specifications

All measurements are given in in.

Year	Engine No.Cyl Displace- ment (cu in.)	CRANKSHAFT				CONNECTING ROD		
		Main Brg. Journal Dia	Main Brg. Oil Clearance	Shaft End-Play	Thrust on No.	Journal Diameter	Oil Clearance	Side Clearance
'68–'71	6-225	2.7495–2.7505	.0005–.0015	.002–.007	3	2.1865–2.1875	.0005–.0015	.006–.012
'68	8-318	2.4495–2.5005	.0005–.0015	.002–.007	3	2.124–2.125	.0005–.0025	.006–.014
'69–'74	8-318	2.4495–2.5005	.0005–.0015	.002–.007	3	2.124–2.125	.0005–.0025	.009–.017
'71–'74	8-360	2.8095–2.8105	.0005–.0025	.002–.007	3	2.124–2.125	.0005–.0015	.009–.017
'67–'74	8-383, 400	2.6245–2.6255	.0005–.0015	.002–.007	3	2.3740–2.3750	.0005–.0015	.009–.017
'67–'74	8-440	2.7495–2.7505	.0005–.0015	.002–.007	3	2.3740–2.3750	.001–.0020	.009–.017
'75–'76	8-318	2.4495–2.5005	.0005–.0025	.002–.010	3	2.124–2.125	.0005–.0025	.010–.018
	8-360	2.8095–2.8105	.0005–.0025	.002–.010	3	2.124–2.125	.0005–.0030	.010–.015
	8-400	2.6245–2.6255	.0005–.0025	.002–.010	3	2.3740–2.3750	.0005–.0025	.009–.017
	8-440	2.7495–2.7505	.0005–.0025	.002–.010	3	2.3740–2.3750	.0005–.0025	.009–.017

Ring Gap

Year	Engine	Top Compression	Bottom Compression	Year	Engine	Oil Control
'68–'72	6-225	.010–.020	.010–.020	'68–'74	All engines	.015–.055
	8-318, 360, 383, 400, 440	.013–.023	.013–.023	'75–'76	All engines	Not applicable
'73–'76	8-318, 360	.010–.020	.010–.020			
'73–'76	8-400, 440	.013–.023	.013–.023			

Ring Side Clearance

Year	Engine	Top Compression	Bottom Compression	Year	Engine	Oil Control
'68–'76	All engines	.0015–.0030	.0015–.0030	'68–'76	6-225, 318, 360	.0002–.005
					8-383, 440, 400	.0000–.005

318 cu in. engine answers the need for a small, reliable, economy powerplant, while the 360 is chosen where more power is desired.

Chrysler Corporation's B block line is really two series of engines, the low-block and high-block series. These sub-series differ in block deck height, main journal diameter, connecting rod length, and pushrod length. Otherwise, they are similar and many parts interchange. The 383 and 400 cu in. engines are low-block types, while the 440 cu in. V8 is of raised-block construction. All these engines are conventional V8s with wedge-shaped combustion chambers and deep blocks that extend well below the crankshaft centerline.

Piston Clearance

Year	Engine	Piston to Bore Clearance (in.)
'68–'76	6-225	.0005–.0015
	8-318, 360	.0005–.0015
	8-383, 400, 440	.0003–.0013

Engine Removal and Installation

WITH TRANSMISSION

1. Scribe the outline of the hood hinge brackets on the bottom of the hood and remove the hood.
2. Drain the cooling system and remove the radiator (see "Engine Cooling" below).
3. Remove the battery (see "Engine Electrical," above).
4. Remove the fuel line from the fuel pump and plug the line.
5. Remove all wires and hoses that attach to the engine. Remove all emission control equipment that may be damaged by the engine removal procedure.
6. If equipped with air conditioning and/or power steering, remove the unit from the engine and position it out of the way *without disconnecting the lines*.
7. Attach lifting sling to the engine. On V8 models remove the carburetor and attach the engine lifting fixture to the carburetor flange studs on the intake manifold. On sixes, attach the sling to the cylinder head.
8. Raise the vehicle on a hoist and install an engine support fixture to support the rear of the engine.
9. On automatic transmission models, drain the transmission and torque converter. On standard transmission models, disconnect the clutch torque shaft from the engine.
10. Disconnect the exhaust pipe/s from the exhaust manifold/s.
11. Remove the driveshaft.
12. Disconnect the transmission linkage and any wiring or cables that attach to the transmission.
13. Remove the engine rear support crossmember and remove the transmission.
14. Remove the bolts that attach the motor mounts to the chassis.
15. Lower the vehicle and attach a chain hoist or other lifting device to the engine.
16. Raise the engine and carefully remove it from the vehicle.

Torque Specifications
All readings in ft lbs

Year	Engine Displacement (cu in.)	Cylinder Head Bolts	Rod Bearing Bolts	Main Bearing Bolts	Crankshaft Pulley Bolt	Flywheel to Crankshaft Bolts	MANIFOLD Intake	MANIFOLD Exhaust
'68–'71	6-All	70	45	85	Press fit	55	10①	10
'68–'76	8-318, 360	95	45	85	135②	65	40	30
	8-383, 400, 440	70	45	85	135②	55	40	30

① Intake to exhaust bolts—20 ft. lbs.
② '71 318, 360 cu in. engines—100 ft lbs.

17. Reverse above procedure to install.

WITHOUT TRANSMISSION

CAUTION: *If the engine is to be removed from the car without the transmission, the weight of the engine must never be allowed to rest on the transmission input shaft (manual) or the torque converter hub (automatic).*

1. Perform Steps 1–7 and 10 of the above removal procedure.

2. If the car is equipped with an automatic transmission, attach a remote starter switch to the engine and remove the bellhousing. Crank the engine to gain access to the converter-to-drive plate attaching nuts and remove the nuts. Remove the bolt that attaches the transmission filler tube to the engine.

3. If the car is equipped with a manual transmission, disconnect the clutch torque shaft from the engine block and the clutch linkage from the adjustment rod. Make marks at twelve o'clock on the flywheel and on the engine to aid in alignment during assembly.

4. Support the transmission and remove the bolts that attach it to the engine or clutch bellhousing. When removing the engine, place a block of wood on the lifting point of a floor jack and position the jack under the transmission. As the engine is removed, raise and lower the jack as required, so that the angle of the transmission duplicates that of the engine as nearly as possible.

CAUTION: *Do not allow the engine to rest on the input shaft or torque converter hub.*

Installation is in the reverse order of removal, however, there are some important points to note:

1. When installing the engine in a car with an automatic transmission, remember that the crankshaft flange bolt circle, the inner and outer circle of holes in the driveplate, and the tapped holes in the face of the converter all have one hole offset. To ensure proper engine/torque converter balance, the torque converter must be mounted on the driveplate in the *same* location in which it was originally installed.

2. When installing the engine in a vehicle with a manual transmission, it may be necessary to disconnect the drive-shaft and turn the transmission (in gear) to get the input shaft splines to mesh with the inner hub on the clutch disc. Remember to use the aligning marks on the flywheel and engine.

Cylinder Head

REMOVAL AND INSTALLATION

6 Cylinder Models

1. Disconnect the battery.

2. The entire cooling system must be drained by opening the drain cock in the radiator and removing the drain plug on the side of the engine block.

3. Remove the vacuum line at the carburetor and distributor.

4. Disconnect the accelerator linkage.

5. Remove the spark plug wires at the plug.

6. Taking note of their positions, disconnect the heater hoses.

7. Disconnect the temperature gauge sending wire.

8. Disconnect the exhaust pipe at the exhaust manifold flange.

9. Disconnect the diverter valve vacuum at the intake manifold and take the air tube assembly from the cylinder head (if so equipped).

10. Remove the PCV and evaporative control system (if so equipped).

11. Remove the intake/exhaust manifold and carburetor as an assembly. Remove the valve cover.

12. Remove the rocker arms and shaft assembly.

13. Remove the pushrods, being sure to take note of their location so they may be installed in their original location.

14. Remove 14 head bolts and the cylinder head.

15. Installation is the reverse of the above. While the head is off, check it for warpage with a straightedge. In addition, install the head gasket with a good quality sealer and be sure to torque the head bolts in the proper sequence to the specified torque. Torque the bolts in three stages.

V8 Engines

1. Drain the cooling system and disconnect the battery ground cable.

2. Remove the alternator, air cleaner, and fuel line.

3. Disconnect the accelerator linkage.

4. Remove the vacuum advance line which runs from the carburetor (or emission control device) to the distributor.

5. Remove the distributor cap and wires as an assembly.

6. Disconnect the coil wires, water temperature sending unit, heater hoses, and by-pass hose. Disconnect the air injection system hoses and components from the cylinder head (if so equipped).

7. Remove the PCV system, the evaporative control system (if so equipped), and the valve covers.

8. Remove the intake manifold, ignition coil, and carburetor as an assembly. On 383, 400, and 440 cu in. V8s, remove the tappet chamber cover.

9. Remove the exhaust manifolds.

10. Remove the rocker and shaft assemblies.

11. Remove the pushrods and keep them in order to ensure installation in their original locations.

12. Remove the head bolts from each cylinder head and remove the cylinder heads.

To install the cylinder heads, proceed as follows:

1. Clean all the gasket surfaces of the engine block and the cylinder heads. Install the spark plugs.

2. Coat new cylinder head gaskets with sealer, install the gaskets, and refit the cylinder heads.

3. Install the cylinder head bolts, and tighten them, in the sequence illustrated, initially to one of the following:

50 ft lbs—318 and 360 cu in. V8s

40 ft lbs—383, 400, and 440 cu in. V8s

4. Continue torquing the bolts in the proper sequence, to the correct figure specified in the "Torque Specifications" chart.

5. Reverse the removal procedure, Steps 1–12, in order to complete installation.

Torque Sequences

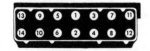

Slant six

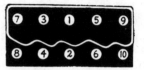

318 and 360 cu in. V8s

383, 400, and 440 cu in. V8s

VALVE GUIDES

Plymouth engines do not have separate valve guides. They do have, however, 0.005, 0.015, and 0.030 in. oversize valves (stem diameter). To use these, ream the worn guides to the smallest oversize that will clean up wear. Always start with the smallest reamer and proceed in steps to the largest, as this maintains the concentricity of the guide with the valve seat.

As an alternate procedure, some local automotive machine shops bore out the stock guides and replace them with bronze or cast iron guides which are of stock internal dimensions.

OVERHAUL

See "Engine Rebuilding" at the end of this chapter for cylinder head overhaul procedures.

Rocker Shafts

REMOVAL AND INSTALLATION

6 Cylinder

The rocker arm shaft has 12 straight steel rocker arms arranged on it with hardened steel spacers fitted between

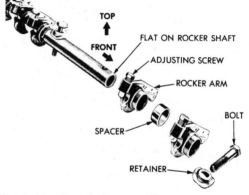

Slant six rocker shaft assembly

each pair of rocker arms. The shaft is secured by bolts and steel retainers which are attached to the 7 cylinder head brackets. To remove the rocker arm and shaft assembly:

1. Remove the PCV system.

2. Remove the evaporative control system (if so equipped).

3. Remove the valve cover and its gasket.

4. Remove the rocker shaft bolts and retainers.

5. Remove the rocker arm and shaft assembly.

6. Reverse the above for installation. The flat on the end of the shaft must be on top and point toward the front of the engine to provide proper lubrication to the rocker arms.

7. Torque all bolts to 25 ft lbs. Before replacing the valve cover, adjust the valves.

V8 Engines

The stamped steel rocker arms are arranged on one rocker arm shaft per cylinder head. Because the angle of the pushrods tends to force the rocker arm pairs toward each other, oilite spacers are fitted to absorb the side thrust at each rocker arm. The shaft is secured by bolts and steel retainers attached to the five brackets on the cylinder head. To remove the arm and shaft from each cylinder head:

1. Disconnect the spark plug wires.

2. Disconnect the PCV system and evaporative control system (if so equipped) from the valve cover.

3. Remove each valve cover and gasket.

4. Remove the rocker shaft bolts and retainer.

5. Remove each rocker arm and shaft assembly.

6. Reverse the above for installation. Be sure to observe the following:

 a. On 318 and 360 cu in. engines, the notch on the end of both rocker shafts should point to the engine centerline, and toward the *front* of the engine on the left cylinder bank and toward the *rear* on the right bank. Torque the rocker shaft mounting bolts to 210 in. lbs.

 b. On 383, 400, and 440 cu in. engines, install the rocker shafts so that the 3/16 in. squirt holes point downward into the rocker arms, and so that the 15° angle of these holes point outward to the valve end of the rocker arms. Torque the rocker shaft bolts to 25 ft lbs.

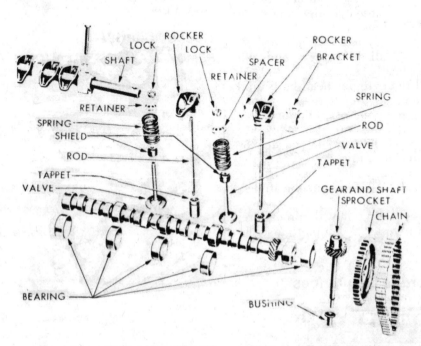

Valve train—383, 400, and 440 cu in. V8s

Intake Manifold

REMOVAL AND INSTALLATION

V8 Engines

1. Drain the cooling system and disconnect the battery.

2. Remove the air cleaner, emission system hoses, vacuum lines, and fuel line from the carburetor.

3. Disconnect the accelerator linkage.

4. Remove the vacuum control between the carburetor and distributor (or emission control system).

5. Remove the distributor cap and wires as a unit.

6. Disconnect the coil wires, temperature sending unit wires, and heater and by-pass hoses.

7. Remove the intake manifold securing bolts and remove the manifold and carburetor as an assembly.

8. To install the manifold, reverse the removal procedure. Be sure to torque the manifold in three steps and remember to use a good commercial sealer on new manifold gaskets. Torque the bolts in the sequence illustrated and to the figure given in the "Torque Specifications" chart.

NOTE: *On 360 cu in. engines, do not use sealer on the composition side gaskets.*

Torque Sequence

V8 engines

Exhaust Manifold

REMOVAL AND INSTALLATION

318 and 360 V8s

1. Remove the exhaust pipe-to-manifold bolts and, on the left side, remove the attaching brace bolt.

2. Remove the manifold-to-cylinder head bolts and remove the manifold.

3. Reinstall by reversing the above. Be sure that the mating surfaces are clean, in complete alignment, and that the heat control valve is free.

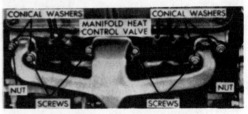

Installation of the V8 exhaust manifold

CAUTION: *The studs must be replaced with new ones if they came out with the nuts. To prevent water leaks, apply sealer on the coarse thread ends of the new studs.*

383, 400, AND 440 CU IN. V8s

1. Disconnect the negative $(-)$ battery cable.

2. Remove the spark plugs, being sure to label their leads to aid in correct installation.

3. Remove the alternator. See "Engine Electrical", above.

4. On models equipped with air injection remove the injection tubes and gaskets (See chapter 4).

5. Unfasten the downpipes from the exhaust manifold flanges.

6. Unfasten the stud nuts which secure the exhaust manifold to the heads and slide the manifolds off the studs.

7. Clean the manifold with solvent and blow dry. Check the manifolds for cracks and distortion.

8. Installation is performed in the reverse order of removal. Use new manifold-to-head gaskets and manifold flange-to-downpipe gaskets. Tighten the manifold nuts in three stages to the figure given in the "Torque Specifications" chart.

CAUTION: *The studs must be replaced with new ones if they come out with the nuts. Apply sealer to the coarse thread ends of the studs.*

Combination Manifold

REMOVAL AND INSTALLATION

6 Cylinder

1. Remove the air cleaner and the fuel line from the carburetor.

Combination manifold used on slant sixes

2. Disconnect the accelerator linkage.

3. Disconnect the exhaust pipe at the exhaust manifold flange.

4. Withdraw the manifold assembly-to-cylinder head bolts and remove the intake and exhaust manifolds with the carburetor as a single unit. The manifolds may be separated by removing the three bolts which hold them together.

Installation is in the following order:

1. Place a new gasket between the intake and exhaust manifolds and install three attaching bolts loosely.

2. With a new gasket in place, position the complete manifold combination on the cylinder head.

3. Install conical washers (cupped side away from the nut) and the nuts. Torque alternately to a final 10 ft lbs.

4. Connect the exhaust pipe to the manifold flange and torque these two bolts to 30 ft lbs.

6. Install the carburetor and connect the fuel line, vacuum line, and the throttle linkage.

7. Install the air cleaner. Start the engine and check for intake and exhaust leaks.

Timing Chain Cover

REMOVAL AND INSTALLATION

6 Cylinder

To remove the timing chain cover:

1. Drain the cooling system completely.

2. Remove the radiator and the fan.

3. Using a suitable puller, remove the vibration damper from the end of the crankshaft.

4. Loosen the oil pan bolts to allow clearance and remove the timing chain cover and its gasket.

To install the timing chain cover:

1. Be sure that the mating surfaces of the cover and the engine block are clean and free from burrs.

2. Using a new gasket, slide the cover over its locating dowels and torque the securing bolts to 15 ft lbs.

3. Be sure that all the oil pan gaskets are in place and tighten the pan bolts to 17 ft lbs.

4. Place the damper pulley assembly hub key in the slot in the crankshaft and lubricate the lip of the oil seal. Slide the hub on the crankshaft.

5. Press the damper pulley assembly onto the crankshaft.

6. Install both the fan and the radiator. Fill up the cooling system.

V8 Engines

To remove the timing chain cover:

1. Drain the cooling system and remove the radiator, fan belt, and water pump assembly.

2. Remove the pulley from the vibration damper. Remove the bolt and washer securing the vibration damper on the crankshaft.

3. Using a puller, remove the vibration damper from the end of the crankshaft.

4. Remove the fuel lines and the fuel pump (318 and 360 only).

5. Loosen the oil pan bolts and remove the front bolt from each side (318 and 360 only).

6. Remove the timing chain cover while being extremely careful not to damage the oil pan gasket.

NOTE: *It is normal to find neoprene particles collected between the crankshaft seal retainer and the crankshaft oil slinger.*

To install the timing chain cover:

1. Be sure that the mating surfaces of the cover and the engine block are clean and free from burrs.

2. Using a new gasket, carefully install the cover to avoid damaging the oil pan gasket.

3. Torque the timing chain cover capscrews to 30 ft lbs on 318 and 360 engines; and to 15 ft lbs on 383, 400, and 440 engines. Be sure that all of the oil pan gaskets are in place and torque the pan capscrews to 200 in. lbs (318 and 360 only).

4. Lubricate the oil seal lip, position the damper hub slot on the key in the

crankshaft, and slide the hub on the crankshaft.

5. Press the damper hub onto the crankshaft.

6. Slide the pulley over the shaft and secure it with the bolts and lockwashers. Torque the bolts to the specifications given in the "Torque Specifications" chart.

7. Install the damper hub retainer washer and bolt.

8. Install the fuel pump and fuel lines.

9. Using new gaskets, install the water pump securing bolts to 30 ft lbs.

10. Install the radiator, fan and belt, and the hoses. Fill the cooling system.

TIMING CHAIN COVER OIL SEAL REPLACEMENT

All Engines

The timing chain cover oil seal can be removed and replaced easily by using a special factory tool. Therefore, it is recommended that the tool be used in this procedure.

1. Remove the timing chain cover.

2. Position the remover screw of the special tool through the timing gear cover with the inside of the cover up.

3. Position the remover blocks directly opposite each other and force the angular lip between the neoprene and the flange of the seal retainer.

4. Place the washer and nut on the remover screw. Tighten the nut, forcing the blocks into the gap until the seal retainer lip is distorted. This will position the remover.

5. Place the sleeve over the retainer.

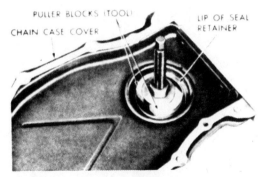

Special seal remover tool with the puller blocks expanded

Then place the removing and installing plate into the sleeve.

6. Fit the flat washer and nut on the remover screw. Hold the center screw and tighten the locknut to remove the seal.

7. Insert the remover screw through the removing and installing plate so that the thin shoulder will be facing up.

8. Position the remover screw and plate through the seal opening with the inside of the cover up.

9. Place the new oil seal in the cover opening with the neoprene down. Place the seal installing plate into the new seal with the protective recess toward the seal retainer lip.

10. Fit the flat washer and nut on the remover screw. Hold the center screw and tighten the locknut to install the new seal.

11. The seal is correctly installed when the neoprene is tight against the face of the cover. Try to insert a 0.0015 in. feeler gauge between the neoprene and the cover. If the seal is installed properly, the feeler gauge cannot be inserted.

CAUTION: *Do not overcompress the neoprene.*

NOTE: *If the special factory tool is unavailable, use extreme care in the removal and installation of the oil seal.*

Timing Chain
CHECKING THE TENSION

Because there is no timing chain tensioner on these engines, the timing chain should be replaced if it is stretched to the point where camshaft sprocket axial motion, with the crankshaft stationary, exceeds $3/16$ in. To check the timing chain stretch, proceed as follows:

1. Remove the timing gear cover.

2. Slide the crankshaft oil slinger off the crankshaft end.

3. Place a straightedge, calibrated in inches, next to the timing chain at the camshaft sprocket. Any chain movement can be measured here.

4. To take up the timing chain slack, use a torque wrench to rotate the camshaft sprocket lockbolt in the direction of crankshaft rotation. Apply a torque of 30

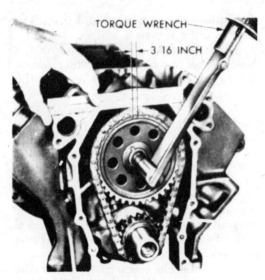

Measuring the timing chain tension

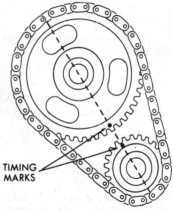

Slant six timing mark alignment

ft lbs with the cylinder heads installed or 15 ft lbs with the heads removed.

NOTE: *When applying the torque to the camshaft sprocket bolt, the crankshaft should not be allowed to move. If necessary, block the crankshaft to prevent rotation.*

5. Now apply the specified torque to the camshaft sprocket in the reverse direction of crankshaft rotation and note the amount of chain movement. Hold the straightedge so that its dimensional rule is even with the edge of a chain link. If chain movement exceeds 3/16 in. (1968–72) or 1/8 in. (1973–76), install a new timing chain.

6. If the chain is satisfactory, slide the crankshaft oil slinger over the shaft and up against the sprocket with the flange away from the sprocket.

7. Install the timing gear cover.

REMOVAL AND INSTALLATION

6 Cylinder

1. Remove the timing gear cover and crankshaft oil slinger.

2. Remove the camshaft sprocket lockbolt and remove the timing chain with the camshaft sprocket.

3. Turn the crankshaft to line up the timing mark on the crankshaft sprocket.

4. Install the camshaft sprocket and the timing chain.

5. Line up the timing marks on the sprockets with the centerline of the crankshaft and camshaft.

6. Torque the camshaft sprocket lockbolt to 35 ft lbs.

7. Slide the crankshaft oil slinger over the shaft and up against the sprocket with the flange away from the sprocket.

8. Install the timing gear cover.

V8 Models

When installing a timing chain on a V8 engine, have an assistant support the camshaft with a screwdriver to prevent the camshaft from contacting the welch plug in the rear of the engine block. Remove the distributor and the oil pump/distributor drive gear.

Position the screwdriver against the rear of the cam gear and be careful not to damage the cam lobes.

1. Disconnect the battery and drain the cooling system.

2. Remove the vibration damper and

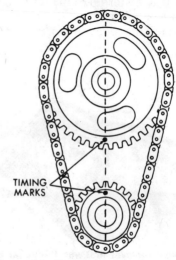

V8 timing mark alignment

pulley. On 318, and 360 cu in. engines, remove the fuel lines and fuel pump, then loosen the oil pan bolts and remove the front bolt on each side.

3. Remove the timing gear cover and the crankshaft oil slinger.

4. On 318, and 360 cu in. engines, remove the camshaft sprocket lockbolt, securing cup washer, and fuel pump eccentric. Remove the timing chain. On 383, 400, and 440 cu in. engines, remove the camshaft sprocket lockbolt and remove the timing chain with the camshaft and crankshaft sprockets.

5. To begin the installation procedure, place the camshaft and crankshaft sprockets on a flat surface with the timing indicators on an imaginary centerline through both sprocket bores. Place the timing chain around both sprockets. Be sure the timing marks are in alignment.

6. Turn the crankshaft and camshaft to align them with the keyway location in the crankshaft sprocket and the dowel hole in the camshaft sprocket.

7. Lift the sprockets and timing chain while keeping the sprockets tight against the chain in the correct position. Slide both sprockets evenly onto their respective shafts.

8. Use a straightedge to measure the alignment of the sprocket timing marks. They must be perfectly aligned.

9. On 318, and 360 cu in. engines, install the fuel pump eccentric, cup washer, and camshaft sprocket lockbolt, and torque to 35 ft lbs. On 383, 400, and 440 V8s, install the washer and camshaft sprocket lockbolt and then torque the lockbolt to 50 ft lbs. Check to make sure that the rear face of the camshaft sprocket is flush with the camshaft end.

Camshaft

The camshafts used on all engines have an integral oil pump and distributor drive gears. The fuel pump eccentric is also integral with the camshaft on all engines except the 318, and 360 cu in. engines, which have a bolt-on fuel pump eccentric. These three engines have a plate on the camshaft to absorb the rearward thrust of the camshaft. The other engines absorb the camshaft thrust on the rear face of the camshaft sprocket hub which bears directly on the front of the engine block.

REMOVAL AND INSTALLATION

When servicing the camshaft, refer to the "Cylinder Head Removal and Installation" procedures because the cylinder head (or heads) must be removed before the camshaft can be removed.

NOTE: *Whenever a new camshaft and/or new tappets are installed, the manufacturer recommends that one quart of their engine oil supplement, or equivalent, should be added to the engine oil to aid break-in. This oil mixture should be left in the engine for a minimum of 500 miles.*

6 Cylinder

1. Remove the cylinder head. Remove the timing gear cover, camshaft sprocket, and timing chain.

2. Remove the valve tappets, keeping them in order to ensure installation in their original location.

3. Remove the crankshaft sprocket.

4. Remove the distributor and the oil pump.

5. Remove the fuel pump.

6. Fit a long bolt into the front of the camshaft to facilitate camshaft removal.

7. Remove the camshaft, being careful not to damage the cam bearings with the cam lobes.

8. Lubricate the camshaft lobes and bearing journals with camshaft lubricant. Insert the camshaft into the engine block.

9. Install the fuel pump and oil pump.

10. Install the distributor (refer to the "Distributor Installation" procedure).

11. Inspect the crowns of all the tappet faces with a straightedge. Replace any tappets that have dished or worn surfaces. Install the tappets.

12. Replace the timing gear, timing gear cover, and the cylinder head.

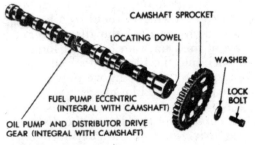

CAMSHAFT SPROCKET

LOCATING DOWEL

WASHER

FUEL PUMP ECCENTRIC
(INTEGRAL WITH CAMSHAFT)

LOCK BOLT

OIL PUMP AND DISTRIBUTOR DRIVE GEAR (INTEGRAL WITH CAMSHAFT)

Camshaft assembly—slant six

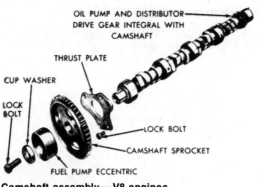

Camshaft assembly—V8 engines

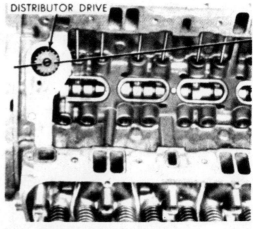

Positioning the distributor drive gear

V8 Engines

NOTE: *On some models, it may be necessary to remove the engine from the car (see "Engine Removal" for correct procedures) and to place it in a suitable stand, before the camshaft can be removed.*

1. Remove the cylinder heads, timing gear cover, sprockets, and the timing chain, as detailed in the sections above.

2. Withdraw the pushrods and the tappets, identifying and keeping them in order, to ensure installation in their original location.

3. Remove the distributor. Extract the oil pump and the distributor driveshaft.

4. Remove the camshaft thrust plate, being careful to note the location of the oil tab (if so equipped).

5. Fit a long bolt into the front of the camshaft and withdraw it, being careful not to damage the camshaft bearings with the cam lobes.

Installation of the camshaft is performed in the following manner:

1. Lubricate the camshaft lobes and bearing journals with a suitable camshaft lubricant. Carefully insert the camshaft so that it is within 2 in. of its final position in the block.

2. Use either the special factory tool or a long screwdriver (held by a helper), inserted in place of the distributor, to prevent the cam from being pushed too far back into the block. This will stop it from pushing out the freeze plug located at the back of the block.

CAUTION: *If a screwdriver is used, be careful not to damage the distributor drive gear or the camshaft lobes with it.*

3. Attach the camshaft thrust plate and the oil tab (if so equipped). Be sure that the oil tab tang enters the hole in the lower right side of the thrust plate. Tighten to 210 in. lbs.

NOTE: *The edge of the tab should be flat against the thrust plate.*

4. Install the oil pump and the distributor driveshaft. Install the distributor, using the above procedure.

5. Inspect the crown of all the tappet faces by using a straightedge. Renew any tappets that have dished or worn surfaces. Install the tappets.

6. Install the timing gear, the timing gear cover, and the cylinder heads as outlined above.

Pistons and Connecting Rods

See the "Engine Rebuilding" section for bearing renewal and ring renewal procedures.

REMOVAL AND INSTALLATION
All Engines

1. Remove the cylinder head and oil pan.

2. Insert a good cylinder ridge reamer into the top of the bores accessible without turning the crankshaft, and remove the ridge. Detach the tool, turn the crankshaft, reattach the tool and remove the ridge on the next cylinder. Continue this process until all cylinder ridges have been removed.

CAUTION: *This is not a boring bar, so merely remove the ridge.*

3. From underneath the car, select the

Piston and Connecting Rod Positioning

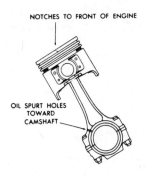

Slant six

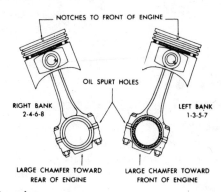

V8 engines

connecting rods in the down position, and remove the locking device (pawl nut or cotter pin). Take off the two nuts that hold the cap to the lower end of the connecting rod. Tap the cap gently and slide it off the end of the bolts. Be careful not to lose the lower half of the rod bearing.

4. Start the connecting rod and piston assembly up toward the top of the bore, but, before pushing it out, replace the cap so that there isn't the slightest chance of it getting mixed up or put on in the wrong way.

5. At this point, note whether the number of the cylinder is stamped on the connecting rod, and, if it is not, some provision will have to be made to mark the rod, such as a file mark or a punch mark. Push the rod and piston assembly up until the rings snap out of the cylinder.

6. When assembling pistons to connecting rods, and the assemblies to the engine, on the slant 6, be sure to locate the squirt hole to the proper side. The 1968–71 engine has the piston head notch at the front, with the oil squirt hole to the right side of the engine.

NOTE: *Late in 1971, a new piston and crankshaft assembly was incorporated into the assembly of some 440 cu. in. non-high-performance engines. Because these pistons are lighter than the old style, they cannot be used as alternative replacements for the older engines. Engines equipped with this new style piston and crank assembly may be identified by the letter "C" on the information pad on the top next to the left bank and tappet rail.*

IDENTIFICATION

All of the pistons are of the same weight, whether they are oversize or not, to maintain correct balance. Oversize pistons are available for honed or rebored cylinder bores in the following sizes: 0.005, 0.020, and 0.040 in.

Piston sizes are stamped on the crown of the piston.

POSITIONING

For all models, the notches go toward the front of the engine during assembly. The oil holes go toward the camshaft, with the exception of the aluminum engine. Look at the following illustrations for correct piston placement.

ENGINE LUBRICATION

Type

6 Cylinder

The engine lubrication system consists of an externally mounted, cam-driven, roto-type pump which is located on the lower right side of the block.

A full-flow disposable oil filter is mounted on the rear of the pump body. Oil is forced from the pump, through the filter, and into a series of internal passages in the engine.

V8 Engines

The lubrication system for the V8 engines is similar to that of the six-cylinder engines, above. The major difference is in the location of the rotor-type oil pump. It is internally mounted on the rear main bearing on 318 and 360 engines. On 400 and 440 engines, the pump is mounted outside the block on the front of the engine.

Oil Pan

REMOVAL AND INSTALLATION

6 Cylinder

1. Raise the vehicle on a hoist.
2. Drain the oil pan.
3. Remove the ball joints from the steering linkage center link.
4. Remove the dust shield.
5. Remove the motor mount stud nuts.
6. Lower the vehicle. Disconnect the battery.
7. Attach an engine lift plate and raise the engine 1½–2 in.
8. Again, raise the vehicle on a hoist. Remove the pan bolts and pan.

Installation is in the reverse order of removal. Tighten the oil pan bolts to 200 in. lbs and the ball joint retaining nuts to 40 ft lbs. Always use a new oil pan gasket.

V8 Engines

1. Disconnect the battery. Remove the dipstick.
2. Raise the vehicle on a hoist. Drain the oil. Remove the engine-to-converter left housing brace.
3. Remove the ball joints from the center steering link.

On 318 and 360 cu in. engines be sure to install the oil pan gasket as shown

4. Remove the crossover pipe from the manifolds.
5. Remove the oil pan bolts and pan.

Installation is the reverse of removal. Be sure that the oil strainer is parallel with the machined surfaces of the block and that it touches the bottom of the oil pan.

Use a new gasket and tighten the oil pan bolts to 200 in. lbs.

REAR MAIN OIL SEAL REPLACEMENT

Service replacement seals are a split-rubber type of composition. This type of seal makes it possible to replace the upper half of the rear main oil seal without removing the engine from the car. When installing rubber seals, they must be replaced as a set and cannot be combined with the rope-type rear main seal. The following procedure is for removing the rope-type seal and replacing it with the rubber-type seal.

1. Remove the oil pan.
2. Remove the rear seal retainer and the rear main bearing cap.
3. Remove the lower rope seal, drive up on either exposed end of the seal with a 6 in. piece of $^3/_{16}$ in. brazing rod. When the opposite end of the seal starts to protrude from the block, have an assistant grasp it with pliers and gently pull it from the block while the opposite end is being driven out.
4. Wipe the crankshaft clean and lightly oil the crankshaft and a new seal before installing the seal.
5. Loosen all main bearing caps slightly to lower the crankshaft which will ease installation.

CAUTION: *Do not allow the crankshaft to drop far enough to permit the main bearings to become displaced on the crankshaft.*

6. Hold the seal tightly against the crankshaft with your thumb (with the paint stripe to the rear) and install the seal in the block groove. Rotate the crankshaft, if necessary, while installing the seal in the groove.

CAUTION: *Make sure the sharp edges on the block groove do not cut or nick the rear of the seal.*

7. Install the lower half of the seal (with the paint stripe to the rear) into the lower seal retainer.
8. Install the rear main bearing cap.

9. Tighten all main bearing caps to specification.

NOTE: *Make sure all main bearings are located in their proper position before tightening the main bearing caps.*

Oil Pump

REMOVAL AND INSTALLATION

6 Cylinder

1. Drain the radiator, disconnect its upper and lower hoses, and remove the fan shroud.
2. Raise the vehicle on a hoist, support the front of the engine with jack stand placed under the right front corner of oil pan, and remove the engine mount bolts. Do not support the engine at the crankshaft pulley or vibration damper.
3. Raise the engine approximately 1½–2 in.
4. Remove the oil filter, oil pump attaching bolts, and pump assembly.

Installation is the reverse of removal. Use a new O-ring, when installing the pump and tighten the oil pump attachment bolts to 200 in. lbs.

318, and 360 V8s

1. Remove the oil pan.
2. Remove the oil pump from the rear main bearing cap.
3. Reverse the above steps to install. Torque the bolts to the proper specifications.

383, 400, and 440 V8s

1. Remove the oil filter.
2. Remove the oil pump attaching bolts.
3. Remove the oil pump.
4. Reverse the above steps to install. Use new O-rings and gaskets.

ENGINE COOLING

Radiator

REMOVAL AND INSTALLATION

1. Drain the cooling system. Detach the oil cooling lines for the automatic transmission if so equipped. Remove the CCEGR valve from the top tank on 1973–76 cars.

2. Disconnect the hose clamps and remove the upper and lower radiator hoses from the radiator.
3. Remove the fan shroud if so equipped. Slide the shroud rearward over the fan and rest it on the engine.
4. Remove the radiator attaching bolts and lift the radiator out of the car.
5. Reverse the above steps to install. On automatic transmission equipped cars, check the fluid level. Fill the cooling system with the proper mixture of glycol antifreeze and water.

Water Pump

REMOVAL AND INSTALLATION

1. Drain the cooling system.
2. Remove the fan shroud securing screws and remove the shroud or position it out of the way (if so equipped).
3. It may be necessary to remove the radiator on some models to gain working space.
4. Remove the alternator belt by loosening the alternator securing bolts. Remove all other accessory drive belts.
5. Remove the fan, spacer or fluid drive, and the water pump pulley. Do not rest the fluid drive with its shaft pointing downward as this will cause the silicon drive fluid to drain down into the fan bearing and cause a bearing failure.
6. On some models, it may be necessary to remove the alternator, air pump, power steering pump, or A/C compressor mounting brackets to gain access to the water pump bolts.

CAUTION: *On cars equipped with air conditioning do not disconnect any refrigerant hoses. Serious injury could result from the high pressure liquid freon contained in these hoses.*

7. Remove the water pump securing bolts and remove the water pump.
8. Install a new water pump on the block using a new gasket and gasket sealer on both sides of the gasket. After tightening the bolts, make sure the pump still rotates. Cars with A/C use a different pump.
9. Refit the alternator and/or compressor brackets if removed.
10. Install the fan, spacer or fluid drive, and pulley.
11. Install all drive belts and tighten to proper tension, ¼ in. deflection at the

mid-point of the longest distance between pulleys.

12. Install the radiator and fan shroud.
Fill the cooling system.

Thermostat

REMOVAL AND INSTALLATION

1. Partially drain the cooling system
to a level slightly below the thermostat.
The thermostat is located in the block
at the end of the upper radiator hose.

2. Remove the upper radiator hose
from the thermostat housing.

3. Remove the thermostat housing
bolts. Remove the housing.

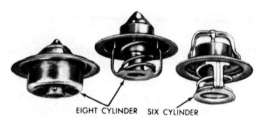

EIGHT CYLINDER   SIX CYLINDER

Install the thermostat with the end at the bottom of the illustration in the block

4. Remove the thermostat from the
block.

5. To install, reverse the above steps.
Use a new gasket and gasket sealer.
Always place the thermostat with the
temperature sensing end facing into the
block.

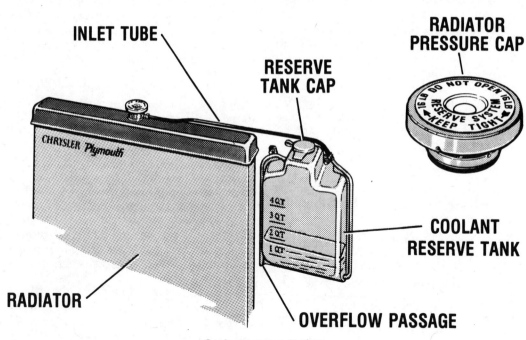

INLET TUBE

RESERVE
TANK CAP

RADIATOR
PRESSURE CAP

CHRYSLER Plymouth

COOLANT
RESERVE TANK

RADIATOR

OVERFLOW PASSAGE

Coolant reserve system

ENGINE REBUILDING

Most procedures involved in rebuilding an engine are fairly standard, regardless of the type of engine involved. This section is a guide accepted rebuilding procedures. Examples of standard rebuilding practices are illustrated and should be used along with specific details concerning your particular engine, found earlier in this chapter.

The procedures given here are those used by any competent rebuilder. Obviously some of the procedures cannot be performed by the do-it-yourself mechanic, but are provided so that you will be familiar with the services that should be offered by rebuilding or machine shops. As an example, in most instances, it is more profitable for the home mechanic to remove the cylinder heads, buy the necessary parts (new valves, seals, keepers, keys, etc.) and deliver these to a machine shop for the necessary work. In this way you will save the money to remove and install the cylinder head and the mark-up on parts.

On the other hand, most of the work involved in rebuilding the lower end is well within the scope of the do-it-yourself mechanic. Only work such as hot-tanking, actually boring the block or Magnafluxing (invisible crack detection) need be sent to a machine shop.

Tools

The tools required for basic engine rebuilding should, with a few exceptions, be those included in a mechanic's tool kit. An accurate torque wrench, and a dial indicator (reading in thousandths) mounted on a universal base should be available. Special tools, where required, are available from the major tool suppliers. The services of a competent automotive machine shop must also be readily available.

Precautions

Aluminum has become increasingly popular for use in engines, due to its low weight and excellent heat transfer characteristics. The following precautions must be observed when handling aluminum (or any other) engine parts:
—Never hot-tank aluminum parts.
—Remove all aluminum parts (identification tags, etc.) from engine parts before hot-tanking (otherwise they will be removed during the process).

—Always coat threads lightly with engine oil or anti-seize compounds before installation, to prevent seizure.
—Never over-torque bolts or spark plugs in aluminum threads. Should stripping occur, threads can be restored using any of a number of thread repair kits available (see next section).

Inspection Techniques

Magnaflux and Zyglo are inspection techniques used to locate material flaws, such as stress cracks. Magnaflux is a magnetic process, applicable only to ferrous materials. The Zyglo process coats the matrial with a fluorescent dye penetrant, and any material may be tested using Zyglo. Specific checks of suspected surface cracks may be made at lower cost and more readily using spot check dye. The dye is sprayed onto the suspected area, wiped off, and the area is then sprayed with a developer. Cracks then will show up brightly.

Overhaul

The section is divided into two parts. The first, Cylinder Head Reconditioning, assumes that the cylinder head is removed from the engine, all manifolds are removed, and the cylinder head is on a workbench. The camshaft should be removed from overhead cam cylinder heads. The second section, Cylinder Block Reconditioning, covers the block, pistons, connecting rods and crankshaft. It is assumed that the engine is mounted on a work stand, and the cylinder head and all accessories are removed.

Procedures are identified as follows:
Unmarked—Basic procedures that must be performed in order to successfully complete the rebuilding process.

Starred (*)—Procedures that should be performed to ensure maximum performance and engine life.

Double starred (**)—Procedures that may be performed to increase engine performance and reliability.

When assembling the engine, any parts that will be in frictional contact must be pre-lubricated, to provide protection on initial start-up. Any product specifically formulated for this purpose may be used. NOTE: *Do not use engine oil.* Where semi-permanent (locked but removable) installation of bolts or nuts is desired, threads should be cleaned and located with Loctite® or a similar product (non-hardening).

Repairing Damaged Threads

Several methods of repairing damaged threads are available. Heli-Coil® (shown here), Keenserts® and Microdot® are among the most widely used. All involve basically the same principle—drilling out stripped threads, tapping the hole and installing a pre-wound insert—making welding, plugging and oversize fasteners unnecessary.

Two types of thread repair inserts are usually supplied—a standard type for most Inch Coarse, Inch Fine, Metric Coarse and Metric Fine thread sizes and a spark plug type to fit most spark plug port sizes. Consult the individual manufacturer's catalog to determine exact applications. Typical thread repair kits will contain a selection of pre-wound threaded inserts, a tap (corresponding to the outside diameter threads of the insert) and an installation tool. Spark plug inserts usually differ because they require a tap equipped with pilot threads and a combined reamer/tap section. Most manufacturers also supply blister-packed thread repair inserts separately in addition to a master kit containing a variety of taps and inserts plus installation tools.

Before effecting a repair to a threaded hole, remove any snapped, broken or damaged bolts or studs. Penetrating oil can be used to free frozen threads; the offending item can be removed with locking pliers or with a screw or stud extractor. After the hole is clear, the thread can be repaired, as follows:

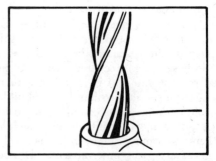

Drill out the damaged threads with specified drill. Drill completely through the hole or to the bottom of a blind hole

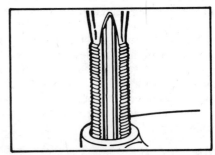

With the tap supplied, tap the hole to receive the thread insert. Keep the tap well oiled and back it out frequently to avoid clogging the threads

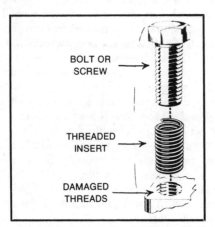

BOLT OR SCREW

THREADED INSERT

DAMAGED THREADS

Damaged bolt holes can be repaired with thread repair inserts

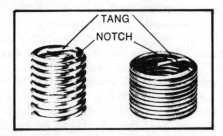

TANG

NOTCH

Standard thread repair insert (left) and spark plug thread insert (right)

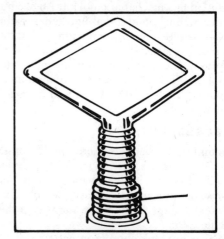

Screw the threaded insert onto the installation tool until the tang engages the slot. Screw the insert into the tapped hole until it is ¼–½ turn below the top surface. After installation break off the tang with a hammer and punch

Standard Torque Specifications and Fastener Markings

The Newton-metre has been designated the world standard for measuring torque and will gradually replace the foot-pound and kilogram-meter. In the absence of specific torques, the following chart can be used as a guide to the maximum safe torque of a particular size/grade of fastener.

- There is no torque difference for fine or coarse threads.
- Torque values are based on clean, dry threads. Reduce the value by 10% if threads are oiled prior to assembly.
- The torque required for aluminum components or fasteners is considerably less.

U. S. BOLTS

SAE Grade Number	1 or 2			5			6 or 7		

Bolt Markings

Manufacturer's marks may vary—number of lines always 2 less than the grade number.

Usage	Frequent			Frequent			Infrequent		
Bolt Size (inches)—(Thread)	Maximum Torque			Maximum Torque			Maximum Torque		
	Ft-Lb	kgm	Nm	Ft-Lb	kgm	Nm	Ft-Lb	kgm	Nm
¼—20	5	0.7	6.8	8	1.1	10.8	10	1.4	13.5
—28	6	0.8	8.1	10	1.4	13.6			
5⁄16—18	11	1.5	14.9	17	2.3	23.0	19	2.6	25.8
—24	13	1.8	17.6	19	2.6	25.7			
⅜—16	18	2.5	24.4	31	4.3	42.0	34	4.7	46.0
—24	20	2.75	27.1	35	4.8	47.5			
7⁄16—14	28	3.8	37.0	49	6.8	66.4	55	7.6	74.5
—20	30	4.2	40.7	55	7.6	74.5			
½—13	39	5.4	52.8	75	10.4	101.7	85	11.75	115.2
—20	41	5.7	55.6	85	11.7	115.2			
9⁄16—12	51	7.0	69.2	110	15.2	149.1	120	16.6	162.7
—18	55	7.6	74.5	120	16.6	162.7			
⅝—11	83	11.5	112.5	150	20.7	203.3	167	23.0	226.5
—18	95	13.1	128.8	170	23.5	230.5			
¾—10	105	14.5	142.3	270	37.3	366.0	280	38.7	379.6
—16	115	15.9	155.9	295	40.8	400.0			
⅞— 9	160	22.1	216.9	395	54.6	535.5	440	60.9	596.5
—14	175	24.2	237.2	435	60.1	589.7			
1— 8	236	32.5	318.6	590	81.6	799.9	660	91.3	894.8
—14	250	34.6	338.9	660	91.3	849.8			

METRIC BOLTS

NOTE: *Metric bolts are marked with a number indicating the relative strength of the bolt. These numbers have nothing to do with size.*

Description	Torque ft-lbs (Nm)			
Thread size x pitch (mm)	Head mark—4		Head mark—7	
6 x 1.0	2.2–2.9	(3.0–3.9)	3.6–5.8	(4.9–7.8)
8 x 1.25	5.8–8.7	(7.9–12)	9.4–14	(13–19)
10 x 1.25	12–17	(16–23)	20–29	(27–39)
12 x 1.25	21–32	(29–43)	35–53	(47–72)
14 x 1.5	35–52	(48–70)	57–85	(77–110)
16 x 1.5	51–77	(67–100)	90–120	(130–160)
18 x 1.5	74–110	(100–150)	130–170	(180–230)
20 x 1.5	110–140	(150–190)	190–240	(160–320)
22 x 1.5	150–190	(200–260)	250–320	(340–430)
24 x 1.5	190–240	(260–320)	310–410	(420–550)

NOTE: *This engine rebuilding section is a guide to accepted rebuilding procedures. Typical examples of standard rebuilding procedures are illustrated. Use these procedures along with the detailed instructions earlier in this chapter, concerning your particular engine.*

Cylinder Head Reconditioning

Procedure	Method
Remove the cylinder head:	See the engine service procedures earlier in this chapter for details concerning specific engines.
Identify the valves:	Invert the cylinder head, and number the valve faces front to rear, using a permanent felt-tip marker.
Remove the rocker arms:	Remove the rocker arms with shaft(s) or balls and nuts. Wire the sets of rockers, balls and nuts together, and identify according to the corresponding valve.
Remove the valves and springs:	Using an appropriate valve spring compressor (depending on the configuration of the cylinder head), compress the valve springs. Lift out the keepers with needlenose pliers, release the compressor, and remove the valve, spring, and spring retainer. See the engine service procedures earlier in this chapter for details concerning specific engines.
Check the valve stem-to-guide clearance: 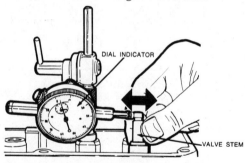 DIAL INDICATOR VALVE STEM **Check the valve stem-to-guide clearance**	Clean the valve stem with lacquer thinner or a similar solvent to remove all gum and varnish. Clean the valve guides using solvent and an expanding wire-type valve guide cleaner. Mount a dial indicator so that the stem is at 90° to the valve stem, as close to the valve guide as possible. Move the valve off its seat, and measure the valve guide-to-stem clearance by rocking the stem back and forth to actuate the dial indicator. Measure the valve stems using a micrometer, and compare to specifications, to determine whether stem or guide wear is responsible for excessive clearance. NOTE: *Consult the Specifications tables earlier in this chapter.*

Cylinder Head Reconditioning

Procedure	Method
De-carbon the cylinder head and valves: 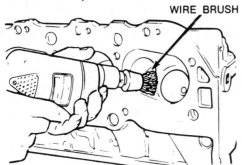 WIRE BRUSH **Remove the carbon from the cylinder head with a wire brush and electric drill**	Chip carbon away from the valve heads, combustion chambers, and ports, using a chisel made of hardwood. Remove the remaining deposits with a stiff wire brush. NOTE: *Be sure that the deposits are actually removed, rather than burnished.*
Hot-tank the cylinder head (cast iron heads only): CAUTION: *Do not hot-tank aluminum parts.*	Have the cylinder head hot-tanked to remove grease, corrosion, and scale from the water passages. NOTE: *In the case of overhead cam cylinder heads, consult the operator to determine whether the camshaft bearings will be damaged by the caustic solution.*
Degrease the remaining cylinder head parts:	Clean the remaining cylinder head parts in an engine cleaning solvent. Do not remove the protective coating from the springs.
Check the cylinder head for warpage: 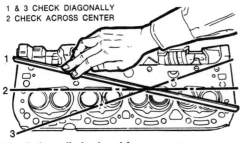 1 & 3 CHECK DIAGONALLY 2 CHECK ACROSS CENTER **Check the cylinder head for warpage**	Place a straight-edge across the gasket surface of the cylinder head. Using feeler gauges, determine the clearance at the center of the straight-edge. If warpage exceeds .003″ in a 6″ span, or .006″ over the total length, the cylinder head must be resurfaced. NOTE: *If warpage exceeds the manufacturer's maximum tolerance for material removal, the cylinder head must be replaced.* When milling the cylinder heads of V-type engines, the intake manifold mounting position is altered, and must be corrected by milling the manifold flange a proportionate amount.
*Knurl the valve guides: 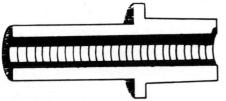 **Cut-away view of a knurled valve guide**	* Valve guides which are not excessively worn or distorted may, in some cases, be knurled rather than replaced. Knurling is a process in which metal is displaced and raised, thereby reducing clearance. Knurling also provides excellent oil control. The possibility of knurling rather than replacing valve guides should be discussed with a machinist.
Replace the valve guides: NOTE: *Valve guides should only be replaced if damaged or if an oversize valve stem is not available.*	See the engine service procedures earlier in this chapter for details concerning specific engines. Depending on the type of cylinder head, valve guides may be pressed, hammered, or shrunk in. In cases where the guides are shrunk into the head, replacement should be left to an equipped machine shop. In other

Cylinder Head Reconditioning

Procedure	Method

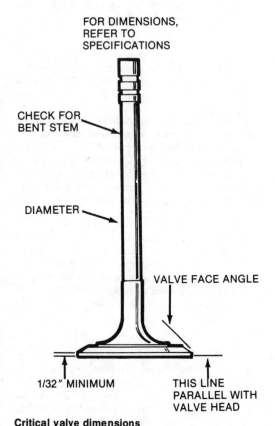

A—VALVE GUIDE I.D.   B—LARGER THAN THE
VALVE GUIDE O.D.

WASHERS

A—VALVE GUIDE I.D.   B—LARGER THAN THE
VALVE GUIDE O.D.

Valve guide installation tool using washers for installation

cases, the guides are replaced using a stepped drift (see illustration). Determine the height above the boss that the guide must extend, and obtain a stack of washers, their I.D. similar to the guide's O.D., of that height. Place the stack of washers on the guide, and insert the guide into the boss.

NOTE: *Valve guides are often tapered or beveled for installation.* Using the stepped installation tool (see illustration), press or tap the guides into position. Ream the guides according to the size of the valve stem.

Replace valve seat inserts:

Replacement of valve seat inserts which are worn beyond resurfacing or broken, if feasible, must be done by a machine shop.

Resurface (grind) the valve face:

Using a valve grinder, resurface the valves according to specifications given earlier in this chapter.

CAUTION: *Valve face angle is not always identical to valve seat angle.* A minimum margin of $1/32''$ should remain after grinding the valve. The valve stem top should also be squared and resurfaced, by placing the stem in the V-block of the grinder, and turning it while pressing lightly against the grinding wheel.

NOTE: *Do not grind sodium filled exhaust valves on a machine. These should be hand lapped.*

FOR DIMENSIONS,
REFER TO
SPECIFICATIONS

CHECK FOR
BENT STEM

DIAMETER

VALVE FACE ANGLE

1/32″ MINIMUM   THIS LINE
PARALLEL WITH
VALVE HEAD

Critical valve dimensions

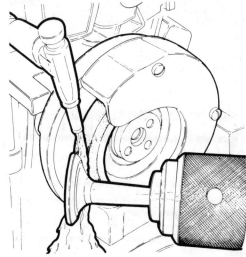

Valve grinding by machine

Cylinder Head Reconditioning

Procedure	Method

Resurface the valve seats using reamers of grinder:

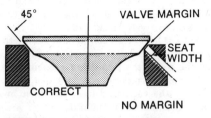

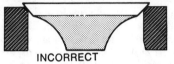

Valve seat width and centering

Reaming the valve seat with a hand reamer

Select a reamer of the correct seat angle, slightly larger than the diameter of the valve seat, and assemble it with a pilot of the correct size. Install the pilot into the valve guide, and using steady pressure, turn the reamer clockwise.

CAUTION: *Do not turn the reamer counterclockwise.* Remove only as much material as necessary to clean the seat. Check the concentricity of the seat (following). If the dye method is not used, coat the valve face with Prussian blue dye, install and rotate it on the valve seat. Using the dye marked area as a centering guide, center and narrow the valve seat to specifications with correction cutters.

NOTE: *When no specifications are available, minimum seat width for exhaust valves should be $5/64''$, intake valves $1/16''$.*

After making correction cuts, check the position of the valve seat on the valve face using Prussian blue dye.

To resurface the seat with a power grinder, select a pilot of the correct size and coarse stone of the proper angle. Lubricate the pilot and move the stone on and off the valve seat at 2 cycles per second, until all flaws are gone. Finish the seat with a fine stone. If necessary the seat can be corrected or narrowed using correction stones.

Check the valve seat concentricity:

Coat the valve face with Prussian blue dye, install the valve, and rotate it on the valve seat. If the entire seat becomes coated, and the valve is known to be concentric, the seat is concentric.

*Install the dial gauge pilot into the guide, and rest of the arm on the valve seat. Zero the gauge, and rotate the arm around the seat. Run-out should not exceed .002″.

Check the valve seat concentricity with a dial gauge

Cylinder Head Reconditioning

Procedure	Method
*Lap the valves: NOTE: *Valve lapping is done to ensure efficient sealing of resurfaced valves and seats.*	Invert the cylinder head, lightly lubricate the valve stems, and install the valves in the head as numbered. Coat valve seats with fine grinding compound, and attach the lapping tool suction cup to a valve head. NOTE: *Moisten the suction cup.* Rotate the tool between the palms, changing position and lifting the tool often to prevent grooving. Lap the valve until a smooth, polished seat is evident. Remove the valve and tool, and rinse away all traces of grinding compound.

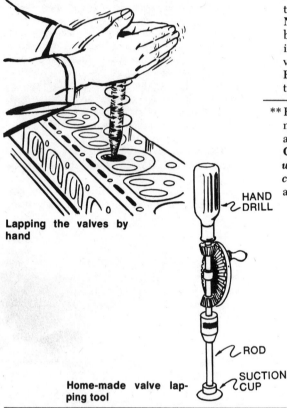

Lapping the valves by hand

Home-made valve lapping tool

Method

** Fasten a suction cup to a piece of drill rod, and mount the rod in a hand drill. Proceed as above, using the hand drill as a lapping tool.
CAUTION: *Due to the higher speeds involved when using the hand drill, care must be exercised to avoid grooving the seat.* Lift the tool and change direction of rotation often.

HAND DRILL

ROD

SUCTION CUP

Check the valve springs:

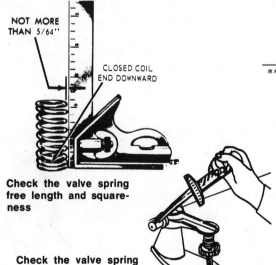

NOT MORE THAN 5/64"

CLOSED COIL END DOWNWARD

Check the valve spring free length and squareness

Check the valve spring test pressure

Method

Place the spring on a flat surface next to a square. Measure the height of the spring, and rotate it against the edge of the square to measure distortion. If spring height varies (by comparison) by more than $1/16''$ or if distortion exceeds $1/16''$, replace the spring.

** In addition to evaluating the spring as above, test the spring pressure at the installed and compressed (installed height minus valve lift) height using a valve spring tester. Springs used on small displacement engines (up to 3 liters) should be $\mp$ 1 lb of all other springs in either position. A tolerance of $\mp$ 5 lbs is permissible on larger engines.

Cylinder Head Reconditioning

Procedure	Method
*Install valve stem seals: 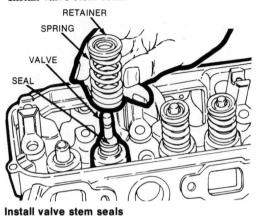 **Install valve stem seals**	* Due to the pressure differential that exists at the ends of the intake valve guides (atmospheric pressure above, manifold vacuum below), oil is drawn through the valve guides into the intake port. This has been alleviated somewhat since the addition of positive crankcase ventilation, which lowers the pressure above the guides. Several types of valve stem seals are available to reduce blow-by. Certain seals simply slip over the stem and guide boss, while others require that the boss be machined. Recently, Teflon guide seals have become popular. Consult a parts supplier or machinist concerning availability and suggested usages. NOTE: *When installing seals, ensure that a small amount of oil is able to pass the seal to lubricate the valve guides; otherwise, excessive wear may result.*
Install the valves:	See the engine service procedures earlier in this chapter for details concerning specific engines. Lubricate the valve stems, and install the valves in the cylinder head as numbered. Lubricate and position the seals (if used) and the valve springs. Install the spring retainers, compress the springs, and insert the keys using needlenose pliers or a tool designed for this purpose. NOTE: *Retain the keys with wheel bearing grease during installation.*
Check valve spring installed height: 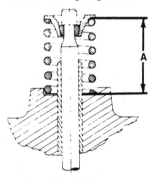 **Valve spring installed height (A)** 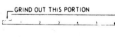  **Measure the valve spring installed height (A) with a modified steel rule**	Measure the distance between the spring pad the lower edge of the spring retainer, and compare to specifications. If the installed height is incorrect, add shim washers between the spring pad and the spring. CAUTION: *Use only washers designed for this purpose.*

Cylinder Head Reconditioning

Procedure	Method
Inspect the rocker arms, balls, studs, and nuts: 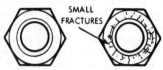 **Stress cracks in the rocker nuts**	Visually inspect the rocker arms, balls, studs, and nuts for cracks, galling, burning, scoring, or wear. If all parts are intact, liberally lubricate the rocker arms and balls, and install them on the cylinder head. If wear is noted on a rocker arm at the point of valve contact, grind it smooth and square, removing as little material as possible. Replace the rocker arm if excessively worn. If a rocker stud shows signs of wear, it must be replaced (see below). If a rocker nut shows stress cracks, replace it. If an exhaust ball is galled or burned, substitute the intake ball from the same cylinder (if it is intact), and install a new intake ball. NOTE: *Avoid using new rocker balls on exhaust valves.*
Replace rocker studs: 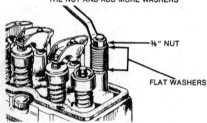 **Extracting a pressed-in rocker stud** 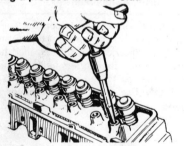 **Ream the stud bore for oversize rocker studs**	In order to remove a threaded stud, lock two nuts on the stud, and unscrew the stud using the lower nut. Coat the lower threads of the new stud with Loctite, and install. Two alternative methods are available for replacing pressed in studs. Remove the damaged stud using a stack of washers and a nut (see illustration). In the first, the boss is reamed .005–.006″ oversize, and an oversize stud pressed in. Control the stud extension over the boss using washers, in the same manner as valve guides. Before installing the stud, coat it with white lead and grease. To retain the stud more positively drill a hole through the stud and boss, and install a roll pin. In the second method, the boss is tapped, and a threaded stud installed.
Inspect the rocker shaft(s) and rocker arms: 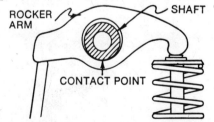 **Check the rocker arm-to-rocker shaft contact area**	Remove the rocker arms, springs and washers from rocker shaft. NOTE: *Lay out parts in the order as they are removed.* Inspect rocker arms for pitting or wear on the valve contact point, or excessive bushing wear. Bushings need only be replaced if wear is excessive, because the rocker arm normally contacts the shaft at one point only. Grind the valve contact point of rocker arm smooth if necessary, removing as little material as possible. If excessive material must be removed to smooth and square the arm, it should be replaced. Clean out all oil holes and passages in rocker shaft. If shaft is grooved or worn, replace it. Lubricate and assemble the rocker shaft.

Cylinder Head Reconditioning

Procedure	Method
Inspect the pushrods:	Remove the pushrods, and, if hollow, clean out the oil passages using fine wire. Roll each pushrod over a piece of clean glass. If a distinct clicking sound is heard as the pushrod rolls, the rod is bent, and must be replaced.
	*The length of all pushrods must be equal. Measure the length of the pushrods, compare to specifications, and replace as necessary.
*Inspect the valve lifters: 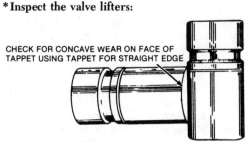 CHECK FOR CONCAVE WEAR ON FACE OF TAPPET USING TAPPET FOR STRAIGHT EDGE **Check the lifter face for squareness**	Remove lifters from their bores, and remove gum and varnish, using solvent. Clean walls of lifter bores. Check lifters for concave wear as illustrated. If face is worn concave, replace lifter, and carefully inspect the camshaft. Lightly lubricate lifter and insert it into its bore. If play is excessive, an oversize lifter must be installed (where possible). Consult a machinist concerning feasibility. If play is satisfactory, remove, lubricate, and reinstall the lifter.
*Testing hydraulic lifter leak down:	Submerge lifter in a container of kerosene. Chuck a used pushrod or its equivalent into a drill press. Position container of kerosene so pushrod acts on the lifter plunger. Pump lifter with the drill press, until resistance increases. Pump several more times to bleed any air out of lifter. Apply very firm, constant pressure to the lifter, and observe rate at which fluid bleeds out of lifter. If the fluid bleeds very quickly (less than 15 seconds), lifter is defective. If the time exceeds 60 seconds, lifter is sticking. In either case, recondition or replace lifter. If lifter is operating properly (leak down time 15–60 seconds), lubricate and install it.

Cylinder Block Reconditioning

Procedure	Method
Checking the main bearing clearance: PLASTIGAGE® **Plastigage® installed on the lower bearing shell**	Invert engine, and remove cap from the bearing to be checked. Using a clean, dry rag, thoroughly clean all oil from crankshaft journal and bearing insert. NOTE: *Plastigage® is soluble in oil; therefore, oil on the journal or bearing could result in erroneous readings.* Place a piece of Plastigage along the full length of journal, reinstall cap, and torque to specifications. NOTE: **Specifications are given in the engine specifications earlier in this chapter.** Remove bearing cap, and determine bearing clearance by comparing width of Plastigage to the scale on Plastigage envelope. Journal taper is determined by comparing width of the Plas-

Cylinder Block Reconditioning

Procedure	Method

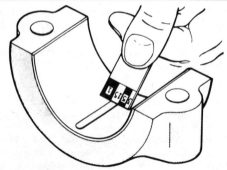

Measure Plastigage® to determine main bearing clearance

tigage strip near its ends. Rotate crankshaft 90° and retest, to determine journal eccentricity. **NOTE:** *Do not rotate crankshaft with Plastigage installed.* If bearing insert and journal appear intact, and are within tolerances, no further main bearing service is required. If bearing or journal appear defective, cause of failure should be determined before replacement.

* Remove crankshaft from block (see below). Measure the main bearing journals at each end twice (90° apart) using a micrometer, to determine diameter, journal taper and eccentricity. If journals are within tolerances, reinstall bearing caps at their specified torque. Using a telescope gauge and micrometer, measure bearing I.D. parallel to piston axis and at 30° on each side of piston axis. Subtract journal O.D. for bearing I.D. to determine oil clearance. If crankshaft journals appear defective, or do not meet tolerances, there is no need to measure bearings; for the crankshaft will require grinding and/or undersize bearings will be required. If bearing appears defective, cause for failure should be determined prior to replacement.

Check the connecting rod bearing clearance:

Connecting rod bearing clearance is checked in the same manner as main bearing clearance, using Plastigage. Before removing the crankshaft, connecting rod side clearance also should be measured and recorded.

* Checking connecting rod bearing clearance, using a micrometer, is identical to checking main bearing clearance. If no other service is required, the piston and rod assemblies need not be removed.

Remove the crankshaft:

Using a punch, mark the corresponding main bearing caps and saddles according to position (i.e., one punch on the front main cap and saddle, two on the second, three on the third, etc.). Using number stamps, identify the corresponding connecting rods and caps, according to cylinder (if no numbers are present). Remove the main and connecting rod caps, and place

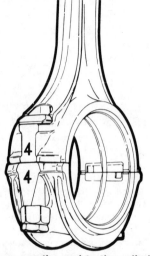

Match the connecting rod to the cylinder with a number stamp

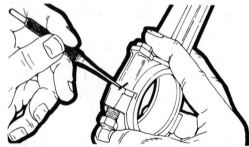

Match the connecting rod and cap with scribe marks

Cylinder Block Reconditioning

Procedure	Method
	sleeves of plastic tubing or vacuum hose over the connecting rod bolts, to protect the journals as the crankshaft is removed. Lift the crankshaft out of the block.
Remove the ridge from the top of the cylinder: RIDGE CAUSED BY CYLINDER WEAR CYLINDER WALL TOP OF PISTON **Cylinder bore ridge**	In order to facilitate removal of the piston and connecting rod, the ridge at the top of the cylinder (unworn area; see illustration) must be removed. Place the piston at the bottom of the bore, and cover it with a rag. Cut the ridge away using a ridge reamer, exercising extreme care to avoid cutting too deeply. Remove the rag, and remove cuttings that remain on the piston. **CAUTION:** *If the ridge is not removed, and new rings are installed, damage to rings will result.*
Remove the piston and connecting rod: **Push the piston out with a hammer handle**	Invert the engine, and push the pistons and connecting rods out of the cylinders. If necessary, tap the connecting rod boss with a wooden hammer handle, to force the piston out. **CAUTION:** *Do not attempt to force the piston past the cylinder ridge* (see above).
Service the crankshaft:	Ensure that all oil holes and passages in the crankshaft are open and free of sludge. If necessary, have the crankshaft ground to the largest possible undersize.
	** Have the crankshaft Magnafluxed, to locate stress cracks. Consult a machinist concerning additional service procedures, such as surface hardening (e.g., nitriding, Tuftriding) to improve wear characteristics, cross drilling and chamfering the oil holes to improve lubrication, and balancing.
Removing freeze plugs:	Drill a small hole in the middle of the freeze plugs. Thread a large sheet metal screw into the hole and remove the plug with a slide hammer.
Remove the oil gallery plugs:	Threaded plugs should be removed using an appropriate (usually square) wrench. To remove soft, pressed in plugs, drill a hole in the plug, and thread in a sheet metal screw. Pull the plug out by the screw using pliers.

Cylinder Block Reconditioning

Procedure	Method
Hot-tank the block: NOTE: *Do not hot-tank aluminum parts.*	Have the block hot-tanked to remove grease, corrosion, and scale from the water jackets. NOTE: *Consult the operator to determine whether the camshaft bearings will be damaged during the hot-tank process.*

Check the block for cracks:	Visually inspect the block for cracks or chips. The most common locations are as follows:
	    Adjacent to freeze plugs.
	    Between the cylinders and water jackets.
	    Adjacent to the main bearing saddles.
	    At the extreme bottom of the cylinders.
	Check only suspected cracks using spot check dye (see introduction). If a crack is located, consult a machinist concerning possible repairs.

	**Magnaflux the block to locate hidden cracks. If cracks are located, consult a machinist about feasibility of repair.
Install the oil gallery plugs and freeze plugs:	Coat freeze plugs with sealer and tap into position using a piece of pipe, slightly smaller than the plug, as a driver. To ensure retention, stake the edges of the plugs. Coat threaded oil gallery plugs with sealer and install. Drive replacement soft plugs into block using a large drift as a driver.
	*Rather than reinstalling lead plugs, drill and tap the holes, and install threaded plugs.
Check the bore diameter and surface: 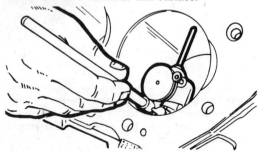 **Measure the cylinder bore with a dial gauge**	Visually inspect the cylinder bores for roughness, scoring, or scuffing. If evident, the cylinder bore must be bored or honed oversize to eliminate imperfections, and the smallest possible oversize piston used. The new pistons should be given to the machinist with the block, so that the cylinders can be bored or honed exactly to the piston size (plus clearance). If no flaws are evident, measure the bore diameter using a telescope gauge and micrometer, or dial gauge, parallel and perpendicular to the engine centerline, at the top (below the ridge) and bottom of the bore. Subtract the bottom measurements from the top to determine taper, and the parallel to

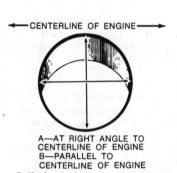

CENTERLINE OF ENGINE

A—AT RIGHT ANGLE TO CENTERLINE OF ENGINE
B—PARALLEL TO CENTERLINE OF ENGINE

Cylinder bore measuring points

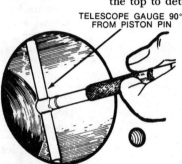

TELESCOPE GAUGE 90° FROM PISTON PIN

Measure the cylinder bore with a telescope gauge

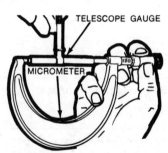

TELESCOPE GAUGE

MICROMETER

Measure the telescope gauge with a micrometer to determine the cylinder bore

Cylinder Block Reconditioning

Procedure	Method
	the centerline measurements from the perpendicular measurements to determine eccentricity. If the measurements are not within specifications, the cylinder must be bored or honed, and an oversize piston installed. If the measurements are within specifications the cylinder may be used as is, with only finish honing (see below). NOTE: *Prior to submitting the block for boring, perform the following operation(s).*
Check the cylinder block bearing alignment:  Check the main bearing saddle alignment	Remove the upper bearing inserts. Place a straightedge in the bearing saddles along the centerline of the crankshaft. If clearance exists between the straightedge and the center saddle, the block must be alignbored.
*****Check the deck height:**	The deck height is the distance from the crankshaft centerline to the block deck. To measure, invert the engine, and install the crankshaft, retaining it with the center main cap. Measure the distance from the crankshaft journal to the block deck, parallel to the cylinder centerline. Measure the diameter of the end (front and rear) main journals, parallel to the centerline of the cylinders, divide the diameter in half, and subtract it from the previous measurement. The results of the front and rear measurements should be identical. If the difference exceeds .005″, the deck height should be corrected. NOTE: *Block deck height and warpage should be corrected at the same time.*
Check the block deck for warpage:	Using a straightedge and feeler gauges, check the block deck for warpage in the same manner that the cylinder head is checked (see Cylinder Head Reconditioning). If warpage exceeds specifications, have the deck resurfaced. NOTE: *In certain cases a specification for total material removal (cylinder head and block deck) is provided. This specification must not be exceeded.*
Clean and inspect the pistons and connecting rods: RING EXPANDER Remove the piston rings	Using a ring expander, remove the rings from the piston. Remove the retaining rings (if so equipped) and remove piston pin. NOTE: *If the piston pin must be pressed out, determine the proper method and use the proper tools; otherwise the piston will distort.* Clean the ring grooves using an appropriate tool, exercising care to avoid cutting too deeply. Thoroughly clean all carbon and varnish from the piston with solvent. CAUTION: *Do not use a wire brush or caustic solvent on pistons.* Inspect the pistons for scuffing, scoring, cracks, pitting, or excessive ring

Cylinder Block Reconditioning

Procedure	Method

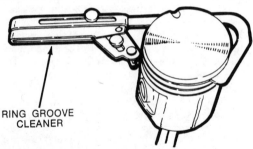

RING GROOVE
CLEANER

Clean the piston ring grooves

groove wear. If wear is evident, the piston must be replaced. Check the connecting rod length by measuring the rod from the inside of the large end to the inside of the small end using calipers (see illustration). All connecting rods should be equal length. Replace any rod that differs from the others in the engine.

* Have the connecting rod alignment checked in an alignment fixture by a machinist. Replace any twisted or bent rods.

* Magnaflux the connecting rods to locate stress cracks. If cracks are found, replace the connecting rod.

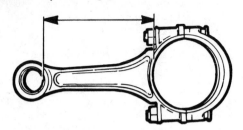

Check the connecting rod length (arrow)

Fit the pistons to the cylinders:

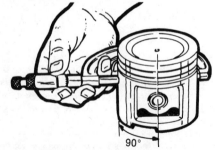

90°

Measure the piston prior to fitting

Using a telescope gauge and micrometer, or a dial gauge, measure the cylinder bore diameter perpendicular to the piston pin, 2½″ below the deck. Measure the piston perpendicular to its pin on the skirt. The difference between the two measurements is the piston clearance. If the clearance is within specifications or slightly below (after boring or honing), finish honing is all that is required. If the clearance is excessive, try to obtain a slightly larger piston to bring clearance within specifications. Where this is not possible, obtain the first oversize piston, and hone (or if necessary, bore) the cylinder to size.

Assemble the pistons and connecting rods:

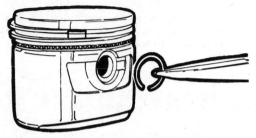

Install the piston pin lock-rings (if used)

Inspect piston pin, connecting rod small end bushing, and piston bore for galling, scoring, or excessive wear. If evident, replace defective part(s). Measure the I.D. of the piston boss and connecting rod small end, and the O.D. of the piston pin. If within specifications, assemble piston pin and rod.
CAUTION: *If piston pin must be pressed in, determine the proper method and use the proper tools; otherwise the piston will distort.*
   Install the lock rings; ensure that they seat properly. If the parts are not within specifications, determine the service method for the type of engine. In some cases, piston and pin are serviced as an assembly when either is defective. Others specify reaming the piston and connecting rods for an oversize pin. If the connecting rod bushing is worn, it may in many cases be replaced. Reaming the piston and replacing the rod bushing are machine shop operations.

Cylinder Block Reconditioning

Procedure	Method

Clean and inspect the camshaft:

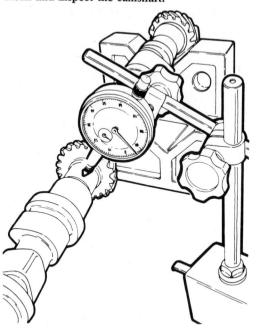

Check the camshaft for straightness

Degrease the camshaft, using solvent, and clean out all oil holes. Visually inspect cam lobes and bearing journals for excessive wear. If a lobe is questionable, check all lobes as indicated below. If a journal or lobe is worn, the camshaft must be regrounded or replaced.

NOTE: *If a journal is worn, there is a good chance that the bushings are worn.* If lobes and journals appear intact, place the front and rear journals in V-blocks, and rest a dial indicator on the center journal. Rotate the camshaft to check straightness. If deviation exceeds .001", replace the camshaft.

* Check the camshaft lobes with a micrometer, by measuring the lobes from the nose to base and again at 90° (see illustration). The lift is determined by subtracting the second measurement from the first. If all exhaust lobes and all intake lobes are not identical, the camshaft must be reground or replaced.

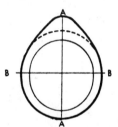

Camshaft lobe measurement

Replace the camshaft bearings:

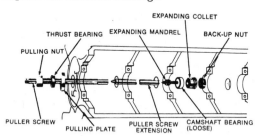

Camshaft bearing removal and installation tool (OHV engines only)

If excessive wear is indicated, or if the engine is being completely rebuilt, camshaft bearings should be replaced as follows: Drive the camshaft rear plug from the block. Assemble the removal puller with its shoulder on the bearing to be removed. Gradually tighten the puller nut until bearing is removed. Remove remaining bearings, leaving the front and rear for last. To remove front and rear bearings, reverse position of the tool, so as to pull the bearings in toward the center of the block. Leave the tool in this position, pilot the new front and rear bearings on the installer, and pull them into position: Return the tool to its original position and pull remaining bearings into position.

NOTE: *Ensure that oil holes align when installing bearings.* Replace camshaft rear plug, and stake it into position to aid retention.

Finish hone the cylinders:

Chuck a flexible drive hone into a power drill, and insert it into the cylinder. Start the hone, and remove it up and down in the cylinder at a rate which will produce approximately a 60° cross-hatch pattern.

NOTE: *Do not extend the hone below the cylinder bore.* After developing the pattern, remove

Cylinder Block Reconditioning

Procedure	Method

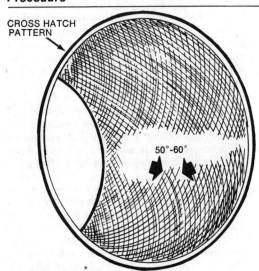

CROSS HATCH PATTERN

50°-60°

Cylinder bore after honing

the hone and recheck piston fit. Wash the cylinders with a detergent and water solution to remove abrasive dust, dry, and wipe several times with a rag soaked in engine oil.

Check piston ring end-gap:

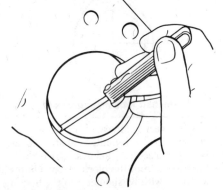

Check the piston ring end gap

Compress the piston rings to be used in a cylinder, one at a time, into that cylinder, and press them approximately 1″ below the deck with an inverted piston. Using feeler gauges, measure the ring end-gap, and compare to specifications. Pull the ring out of the cylinder and file the ends with a fine file to obtain proper clearance.
CAUTION: *If inadequate ring end-gap is utilized, ring breakage will result.*

Install the piston rings:

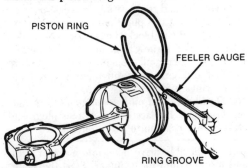

PISTON RING

FEELER GAUGE

RING GROOVE

Check the piston ring side clearance

Inspect the ring grooves in the piston for excessive wear or taper. If necessary, recut the groove(s) for use with an overwidth ring or a standard ring and spacer. If the groove is worn uniformly, overwidth rings, or standard rings and spacers may be installed without recutting. Roll the outside of the ring around the groove to check for burrs or deposits. If any are found, remove with a fine file. Hold the ring in the groove, and measure side clearance. If necessary, correct as indicated above.
NOTE: *Always install any additional spacers above the piston ring.*
The ring groove must be deep enough to allow the ring to seat below the lands (see illustration). In many cases, a "go-no-go" depth gauge will be provided with the piston rings. Shallow grooves may be corrected by recutting, while deep grooves require some type of filler or expander

Cylinder Block Reconditioning

Procedure	Method
	behind the piston. Consult the piston ring supplier concerning the suggested method. Install the rings on the piston, lowest ring first, using a ring expander. NOTE: *Position the rings as specified by the manufacturer.* Consult the engine service procedures earlier in this chapter for details concerning specific engines.
Install the camshaft:	Liberally lubricate the camshaft lobes and journals, and install the camshaft. CAUTION: *Exercise extreme care to avoid damaging the bearings when inserting the camshaft.* Install and tighten the camshaft thrust plate retaining bolts.
	See the engine service procedures earlier in this chapter for details concerning specific engines.
Check camshaft end-play (OHV engines only): **Check the camshaft end-play with a feeler gauge**	Using feeler gauges, determine whether the clearance between the camshaft boss (or gear) and backing plate is within specifications. Install shims behind the thrust plate, or reposition the camshaft gear and retest endplay. In some cases, adjustment is by replacing the thrust plate. See the engine service procedures earlier in this chapter for details concerning specific engines.
DIAL INDICATOR CAMSHAFT **Check the camshaft end-play with a dial indicator**	*Mount a dial indicator stand so that the stem of the dial indicator rests on the nose of the camshaft, parallel to the camshaft axis. Push the camshaft as far in as possible and zero the gauge. Move the camshaft outward to determine the amount of camshaft endplay. If the endplay is not within tolerance, install shims behind the thrust plate, or reposition the camshaft gear and retest. See the engine service procedures earlier in this chapter for details concerning specific engines.
Install the rear main seal:	See the engine service procedures earlier in this chapter for details concerning specific engines.
Install the crankshaft: INSTALLING BEARING SHELL REMOVING BEARING SHELL **Remove or install the upper bearing insert using a roll-out pin**	Thoroughly clean the main bearing saddles and caps. Place the upper halves of the bearing inserts on the saddles and press into position. NOTE: *Ensure that the oil holes align.* Press the corresponding bearing inserts into the main bearing caps. Lubricate the upper main bearings, and lay the crankshaft in position. Place a strip of Plastigage on each of the crankshaft journals, install the main caps, and torque to specifications. Remove the main caps, and compare the Plastigage to the scale on the Plastigage envelope. If clearances are within tolerances, remove the Plastigage, turn the crankshaft 90°, wipe off all oil and retest. If all clearances are correct,

Cylinder Block Reconditioning

Procedure	Method

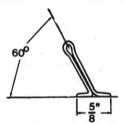

Home-made bearing roll-out pin

remove all Plastigage, thoroughly lubricate the main caps and bearing journals, and install the main caps. If clearances are not within tolerance, the upper bearing inserts may be removed, without removing the crankshaft, using a bearing roll out pin (see illustration). Roll in a bearing that will provide proper clearance, and retest. Torque all main caps, excluding the thrust bearing cap, to specifications. Tighten the thrust bearing cap finger tight. To properly align the thrust bearing, pry the crankshaft the extent of its axial travel several times, the last movement held toward the front of the engine, and torque the thrust bearing cap to specifications. Determine the crankshaft end-play (see below), and bring within tolerance with thrust washers.

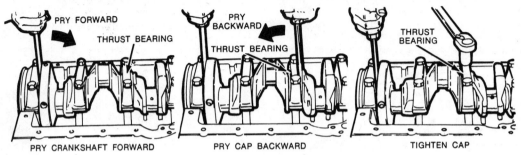

Aligning the thrust bearing

Measure crankshaft end-play:

Mount a dial indicator stand on the front of the block, with the dial indicator stem resting on the nose of the crankshaft, parallel to the crankshaft axis. Pry the crankshaft the extent of its travel rearward, and zero the indicator. Pry the crankshaft forward and record crankshaft end-play. **NOTE:** *Crankshaft end-play also may be measured at the thrust bearing, using feeler gauges (see illustration).*

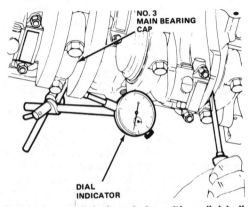

Check the crankshaft end-play with a dial indicator

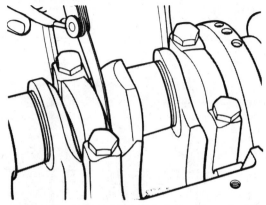

Check the crankshaft end-play with a feeler gauge

Cylinder Block Reconditioning

Procedure	Method
Install the pistons:	Press the upper connecting rod bearing halves into the connecting rods, and the lower halves into the connecting rod caps. Position the piston ring gaps according to specifications (see car section), and lubricate the pistons. Install a ring compresser on a piston, and press two long (8″) pieces of plastic tubing over the rod bolts. Using the tubes as a guide, press the pistons into the bores and onto the crankshaft with a wooden hammer handle. After seating the rod on the crankshaft journal, remove the tubes and install the cap finger tight. Install the remaining pistons in the same manner. Invert the engine and check the bearing clearance at two points (90° apart) on each journal with Plastigage. **NOTE:** *Do not turn the crankshaft with Plastigage installed.* If clearance is within tolerances, remove *all* Plastigage, thoroughly lubricate the journals, and torque the rod caps to specifications. If clearance is not within specifications, install different thickness bearing inserts and recheck. **CAUTION:** *Never shim or file the connecting rods or caps.* Always install plastic tube sleeves over the rod bolts when the caps are not installed, to protect the crankshaft journals.

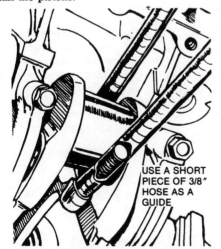

USE A SHORT PIECE OF 3/8″ HOSE AS A GUIDE

Use lengths of vacuum hose or rubber tubing to protect the crankshaft journals and cylinder walls during piston installation

RING COMPRESSOR
Install the piston using a ring compressor

Check connecting rod side clearance:	Determine the clearance between the sides of the connecting rods and the crankshaft using feeler gauges. If clearance is below the minimum tolerance, the rod may be machined to provide adequate clearance. If clearance is excessive, substitute an unworn rod, and recheck. If clearance is still outside specifications, the crankshaft must be welded and reground, or replaced.

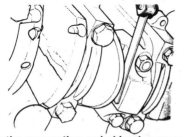

Check the connecting rod side clearance with a feeler gauge

Inspect the timing chain (or belt):	Visually inspect the timing chain for broken or loose links, and replace the chain if any are found. If the chain will flex sideways, it must be replaced. Install the timing chain as specified. Be sure the timing belt is not stretched, frayed or broken. **NOTE:** *If the original timing chain is to be reused, install it in its original position.*

Cylinder Block Reconditioning

Procedure	Method
Check timing gear backlash and runout (OHV engines):	Mount a dial indicator with its stem resting on a tooth of the camshaft gear (as illustrated). Rotate the gear until all slack is removed, and zero the indicator. Rotate the gear in the opposite direction until slack is removed, and record gear backlash. Mount the indicator with its stem resting on the edge of the camshaft gear, parallel to the axis of the camshaft. Zero the indicator, and turn the camshaft gear one full turn, recording the runout. If either backlash or runout exceed specifications, replace the worn gear(s).

Check the camshaft gear backlash

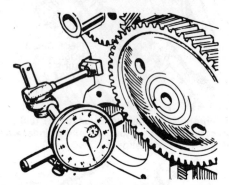

Check the camshaft gear run-out

Completing the Rebuilding Process

Follow the above procedures, complete the rebuilding process as follows:

Fill the oil pump with oil, to prevent cavitating (sucking air) on initial engine start up. Install the oil pump and the pickup tube on the engine. Coat the oil pan gasket as necessary, and install the gasket and the oil pan. Mount the flywheel and the crankshaft vibration damper or pulley on the crankshaft. NOTE: *Always use new bolts when installing the flywheel.* Inspect the clutch shaft pilot bushing in the crankshaft. If the bushing is excessively worn, remove it with an expanding puller and a slide hammer, and tap a new bushing into place.

Position the engine, cylinder head side up. Lubricate the lifters, and install them into their bores. Install the cylinder head, and torque it as specified. Insert the pushrods and install the rocker shaft(s) or position the rocker arms on the pushrods. Adjust the valves.

Install the intake and exhaust manifolds, the carburetor(s), the distributor and spark plugs. Adjust the point gap and the static ignition timing. Mount all accessories and install the engine in the car. Fill the radiator with coolant, and the crankcase with high quality engine oil.

Break-in Procedure

Start the engine, and allow it to run at low speed for a few minutes, while checking for leaks. Stop the engine, check the oil level, and fill as necessary. Restart the engine, and fill the cooling system to capacity. Check the point dwell angle and adjust the ignition timing and the valves. Run the engine at low to medium speed (800–2500 rpm) for approximately ½ hour, and retorque the cylinder head bolts. Road test the car, and check again for leaks.

Follow the manufacturer's recommended engine break-in procedure and maintenance schedule for new engines.

Emission Controls and Fuel System

EMISSION CONTROLS

System Descriptions

POSITIVE CRANKCASE VENTILATION

All models are equipped with a positive crankcase ventilation system which draws air into the engine through the oil filler cap or the air cleaner and circulates it through the engine. The air combines with vapors in the crankcase and exits the engine through a metering valve mounted in the rocker arm cover. The air vapor mixture then re-enters the engine through the carburetor or intake manifold and passes into the combustion chamber where it is burned.

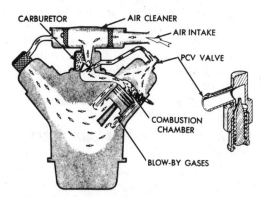

Positive crankcase ventilation (PCV) system

CLEANER AIR PACKAGE (CAP)

All 1968–69 models use this package to reduce engine exhaust emissions. Changes include the addition of limiters to the carburetor idle mixture screws, leaner carburetor mixtures and vacuum controlled ignition timing retard mechanisms.

CLEANER AIR SYSTEM (CAS)

All 1970 and later models are equipped with this type of exhaust emission control. This system consists of: heated carburetor air cleaner intake ducts, carburetor modifications, ignition timing controls, and reduced engine compression ratios.

In addition to the aforementioned controls, many new ones were added or modified since.

Intake Manifold/Cylinder Head Design Change

A change has been incorporated in intake manifold design to place all manifold branch runners on one level. This is in contrast to the previous, two-level design. This change was made to improve fuel vaporization during warm-up, and hence, to allow the use of leaner fuel/air mixtures.

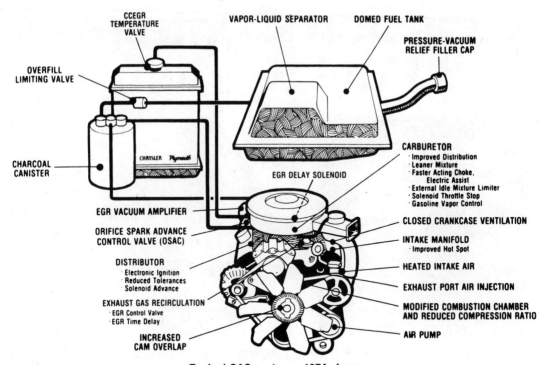

CCEGR
TEMPERATURE
VALVE

VAPOR-LIQUID SEPARATOR

DOMED FUEL TANK

PRESSURE-VACUUM
RELIEF FILLER CAP

OVERFILL
LIMITING VALVE

CHARCOAL
CANISTER

CHRYSLER *Plymouth*

CARBURETOR
· Improved Distribution
· Leaner Mixture
· Faster Acting Choke,
   Electric Assist
· External Idle Mixture Limiter
· Solenoid Throttle Stop
· Gasoline Vapor Control

EGR DELAY SOLENOID

EGR VACUUM AMPLIFIER

ORIFICE SPARK ADVANCE
CONTROL VALVE (OSAC)

DISTRIBUTOR
· Electronic Ignition
· Reduced Tolerances
· Solenoid Advance

EXHAUST GAS RECIRCULATION
· EGR Control Valve
· EGR Time Delay

INCREASED
CAM OVERLAP

CLOSED CRANKCASE VENTILATION

INTAKE MANIFOLD
· Improved Hot Spot

HEATED INTAKE AIR

EXHAUST PORT AIR INJECTION

MODIFIED COMBUSTION CHAMBER
AND REDUCED COMPRESSION RATIO

AIR PUMP

Typical CAS system—1974 shown

Cylinder head design has also been modified. By redesigning the intake ports to give more fuel turbulence and increasing the volume of the combustion chambers to increase quench area, the fuel/air mixture burns more uniformly and this results in a lower production of HC and CO.

Ignition Retard Solenoid 1971 and Earlier Models

The function of this unit is to retard the ignition timing at closed throttle. Located on the distributor side, this solenoid must be operating when the ignition timing is adjusted. To be sure that the solenoid is operating, disconnect the ground lead after the timing is set. If the engine idle speed increases noticeably, the solenoid is functioning properly.

In contrast to the above, disconnect the solenoid when checking the dwell. If this is not done, the dwell meter will not read accurately.

Ignition Advance Solenoid

This solenoid, located on the distributor side, is connected to the starter relay so that it operates only during engine cranking, to improve starting.

When the engine fires, this solenoid ceases to operate. It is designed to advance the timing only during cranking. If the solenoid is not operating, it will affect starting, but not drivability. A possible sign of failure is popping through the carburetor during engine cranking.

VAPOR SAVOR SYSTEM—1972–76

This system is used to prevent the loss of fuel vapor from the fuel tank and carburetor. By venting carburetor and fuel tank vapors to a charcoal canister for temporary storage fuel vapors are prevented from entering the atmosphere. The system is purged of vapors when the engine is running by means of air drawn through the canister by intake manifold vacuum.

In addition, a limiting valve is used to prevent fuel tank overfilling. Located in the fuel vapor vent line in the engine compartment, this valve prevents overfilling by closing when the filler tube is closed by incoming fuel.

AIR INJECTION SYSTEM (AIR PUMP)— 1972–76

In 1972 the air injection system was used on the 225, 400 and 440 cu in. engines sold in California only. In 1973, the 225, 360, and 440 cu in. engines sold

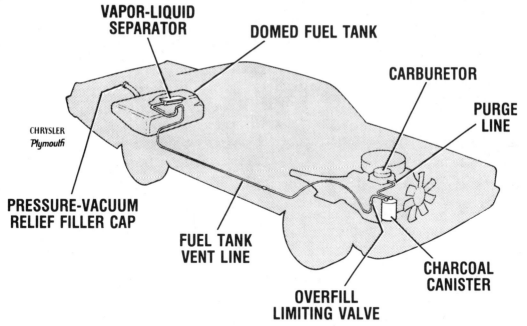

VAPOR-LIQUID SEPARATOR

DOMED FUEL TANK

CARBURETOR

PURGE LINE

CHRYSLER
Plymouth

PRESSURE-VACUUM RELIEF FILLER CAP

FUEL TANK VENT LINE

OVERFILL LIMITING VALVE

CHARCOAL CANISTER

Components of the evaporative emission control system

in California used air injection. In 1974, some of the 400 and 440 cu in. engines sold in California use air pumps, depending upon model usage. Since 1975, all California and most federal engines use air pumps.

A belt-driven air pump, mounted on the front of the engine, is used to inject air into the exhaust ports. This causes oxidation of these gases and a considerable reduction in carbon monoxide and hydrocarbons. The system consists of the pump, a check valve to protect the hoses and pump from hot gases, and a diverter/pressure relief valve assembly.

Service to the air injection system is limited to belt tension adjustment every 12,000 miles (15,000 miles—1975–76). In addition, if any part fails in service, repair is effected by removal and replacement only.

EXHAUST GAS RECIRCULATION

1972—California

In order to reduce the emission of oxides and nitrogen (NO_x), exhaust gases are ducted from the intake manifold crossover passage to contaminate the fuel/air mixture. (Do not confuse this system with other changes in the intake manifold.) These gases are introduced to the intake manifold floor by small jets. Every 12,000 miles, inspect these jets by

looking through the carburetor. If the jet is plugged, remove, clean, and reinstall it.

1973—All

Starting with 1973, all Chrysler Corp. cars use exhaust gas recirculation (EGR). All engines have floor jets like those used on the 1972 models sold in California. The 400 cu in. engine equipped with 4-bbl carburetor uses only floor jets for 1973.

In addition to the floor jets, all other 1973 engines use an EGR control valve. This valve directs exhaust gas from the crossover passage into the intake manifold. By using either ported-vacuum or venturi-vacuum signals, the EGR valve is able to proportion the exhaust gas flow to the amount of vacuum present in the carburetor. Ported-vacuum is used on the 318 cu in. (except California) and the 400 cu in. 2-bbl engines, as well as, the 440 cu in. 4-bbl high-performance engine. Venturi-vacuum is used on the 440 standard engines, and on the 318 2-bbl sold in California.

A thermal switch is used to deenergize the EGR valve when the outside temperature is below 58° F, to provide better driveability.

Starting around 15 March 1973, the ambient temperature sensor was dropped. It has been replaced by a thermostatic valve which is threaded into the top tank of the radiator. A hose runs from one valve nipple to the EGR vacuum amplifier. The other nipple has a filter fitted over it. When the coolant temperature is below 62° F, the valve is opened to the atmosphere, thus preventing the EGR valve diaphragm from getting vacuum. Above 62° F, the valve closes and the EGR valve is allowed to function.

1974-76-All

Floor jets have been dropped from all 1974 engines. The EGR temperature switch (mounted in the upper radiator tank), which was introduced in March 1973, has been retained.

NOTE: *The thermostatic switch for the EGR system is mounted on the thermostat housing on the 360-4 V engine.*

All 1974 engines, except for the following have a vacuum amplifier:
- V8-318—All
- V8-360—4-bbl High Performance
- V8-400—4-bbl High Performance
- V8-440—4-bbl High Performance

All 1975–76 engines use an EGR amplifier, except the 400 Lean Burn engine.

1973—ELECTRIC ASSIST CHOKE

During warm weather a heating element, located in the automatic choke well, comes on to shorten the period of choke operation and thus reduce hydrocarbon emissions. The heating element is operated by a time-delay control switch located next to the choke well. The assist choke draws about three amps of current during operation.

1974—ELECTRIC ASSIST CHOKE

A two-stage electric assist choke is used for 1974. The two-stage choke may be identified by its external resistor:
- Blue resistor 5 ohm—V8-318
- White resistor 10 ohm—All other V8s

Below 58° F, the heating element gets full, low amperage current from the choke control. Above 58° F, the resistor cuts the current in half. After several minutes of operation above 58° F, the control opens the circuit so that the heating element gets no current at all.

Most engines use a 20-watt heating element, except for the following which

have 4-bbl carburetors and use a 40-watt choke:
- V8-440 (Thermo-Quad)—All states
- V8-400 H.P. (Thermo-Quad)—All (except Calif.)
- V8-440 H.P. (Thermo-Quad)—Calif. only
- V8-360—All states

The 40-watt choke has a white paint spot on the choke cover.

1975-76—ELECTRIC ASSIST CHOKE

All 1975 Plymouth engines use a single-stage electric assist choke, similar to that used on 1973 models. 1976 models go back to the dual stage type.

NOₓ SYSTEM

Many 1971 and 1972 vehicles have a NO_x system to control the emission of oxides of nitrogen. Engines with this system all have a special camshaft and a 185° F thermostat.

Manual Transmission

The manual transmission NO_x system uses a transmission switch, a thermal switch, and a solenoid vacuum valve. The transmission switch is screwed into the transmission housing and is closed, except in high gear. The thermal switch, mounted on the firewall, is open whenever the ambient temperature is above 70° F. With the transmission in any gear except high and the temperature above 70°, the solenoid vacuum valve is energized. This shuts off the distributor vacuum advance line preventing vacuum advance. Below 70°, the vacuum advance functions normally.

Automatic Transmission

The NO_x system for automatic transmissions is more complex than the manual transmission system. It prevents vacuum advance when the ambient temperature is above 70° F, speed is below 30 mph, or the car is accelerating. The solenoid vacuum valve is interchangeable with that used in the manual transmission system. The speed switch senses vehicle speed and is driven by the speedometer cable. The control unit is mounted on the firewall. It contains a control module, thermal switch, and a vacuum switch. The control unit senses ambient temperature and manifold vacuum.

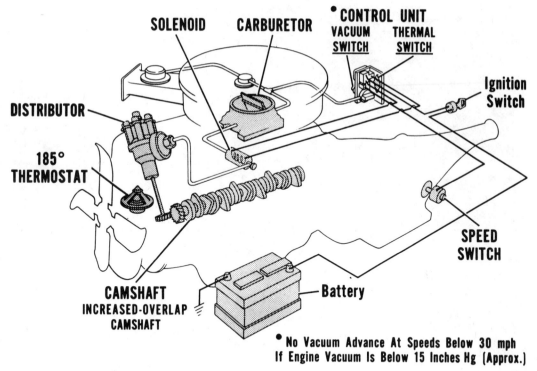

SOLENOID · CARBURETOR · CONTROL UNIT
VACUUM SWITCH · THERMAL SWITCH

Ignition Switch

DISTRIBUTOR

185° THERMOSTAT

SPEED SWITCH

CAMSHAFT
INCREASED-OVERLAP CAMSHAFT

Battery

• No Vacuum Advance At Speeds Below 30 mph
If Engine Vacuum Is Below 15 Inches Hg (Approx.)

1971–72 NO$_x$ control system components

1973—OSAC Valve

Starting with the 1973 models, an orifice spark advance control (OSAC) valve is used to delay distributor vacuum advance for about 15 seconds during acceleration.

NOTE: *The amount of time-delay varies slightly from one engine size to another.*

To aid in cold weather engine operation, a temperature sensing switch is built in to the OSAC valve so that it will not function when the air temperature is below 68 ° F.

Some time after 1 March and before 15 March 1973, the temperature sensor was removed from the OSAC valve, but the general appearance and location of the valve were not changed. The valve can be recognized by a white gasket and a stick-on label with the new part number (3755499).

At the same time, the ignition timing was changed to TDC.

NOTE: *See the engine tune-up specifications decal for further timing information.*

1974–76—OSAC Valve

The OSAC valve has been moved from the firewall to the air cleaner for 1974 and the temperature control restored. There are six different time delay and operating temperature combinations for the valve. These combinations are identified by a color code tape on the top of the valve. These codes are as follows:

Color	Time (sec)	Temperature (°F)
Green	17	58
Red	17	50
Blue	17	①
White	27	58
Orange	27	①
Yellow	27	58

①—No temperature control used

CAUTION: *Always replace the valve with one having the same color code. Failure to do so could result in poor vehicle performance or lack of compliance with the emission laws.*

NOTE: *The 1976 400 Lean Burn engine does not use an OSAC valve.*

1974–76 OSAC valves are mounted on the air cleaner

CATALYTIC CONVERTER—1976–76

All 1975 Chrysler products sold in California, and almost all sold nationwide, are equipped with catalytic converters. These devices are used to oxidize excess carbon monoxide (CO) and hydrocarbons (HC) in the exhaust system before they can escape out the tailpipe and into the atmosphere. The converter is installed in front of the mufflers, underneath the car, and protected by a heat shield.

The expected catalyst life is 50,000 miles, provided that the engine is kept in tune and unleaded fuel is used.

On 1975 models, to keep the catalyst from being overheated by an overly rich mixture during deceleration, a catalyst protection system (CPS) is used. The system consists of a throttle positioner solenoid (not to be confused with the idle stop solenoid), a control box, and an engine rpm sensor.

Any time that the engine speed is more than 2,000 rpm, the solenoid is energized and keeps the throttle butterfly from fully closing, thus preventing the mixture from becoming too rich.

The catalyst protection system is not used on 1976 models.

COOLANT CONTROL IDLE ENRICHMENT (CCIE) SYSTEM—1975–76

The CCIE system is used on 1975–76 Plymouth models equipped with automatic transmissions. The system consists of a vacuum-operated valve built into the carburetor, which shuts off the idle circuit air bleeds when vacuum is supplied to its diaphragm.

Depending upon engine application, vacuum is either routed through a coolant controlled vacuum valve, or a coolant controlled vacuum valve and an EGR vacuum control solenoid.

Vacuum is passed to the valve diaphragm below a predetermined temperature, and on models with an EGR control solenoid, for only 35 seconds after the engine is started. The CCIE valve action closes off the air bleed passages, which richens the mixture, and allows a smoother cold idle.

CHRYSLER'S LEAN BURN SYSTEM—1976 400-4V

This system, new for 1976, is based on the principle that lower NOx emissions would occur if the air/fuel ratio inside the cylinder area was raised from its current point (15.5:1) to a much leaner point (18:1).

In order to make the engine workable, a solution to the problems of carburetion and timing had to be found since a lean running engine is not the most efficient in terms of driveability. Chrysler adapted a conventional Thermo-Quad carburetor to handle the added air coming in, but the real advance of the system is the Spark Control Computer mounted on the air cleaner.

Since a lean burning engine demands precise ignition timing, additional spark control was needed for the distributor. The computer supplies this control by providing an infinitely variable advance curve. Input data is fed instantaneously to the computer by a series of seven sensors located in the engine compartment which monitor timing, water tempera-

ture, air temperature, throttle position, idle/off-idle operation, and intake manifold vacuum. The program schedule module of the spark control computer receives the information from the sensors, processes it, and then directs the ignition control module to advance or retard the timing as necessary. This whole process is going on continuously as the engine is running, taking only a thousandth of a second to complete a circuit from sensor to distributor.

The components of the system are as follows: Modified Thermo-Quad carburetor; Spark Control Computer, consisting of two interacting modules: the Program Schedule Module which is responsible for translating input data, and the Ignition Control Module which transmits data to the distributor to advance or retard the timing; and the following sensors.

Start Pick-up Sensor, located inside the distributor, supplies a signal to the computer providing a fixed timing point which is only used for starting the car. It also has a back-up function of taking over engine timing in case the run pick-up fails. Since the timing in this pick-up is fixed at one point, the engine will be able to run, but not very well.

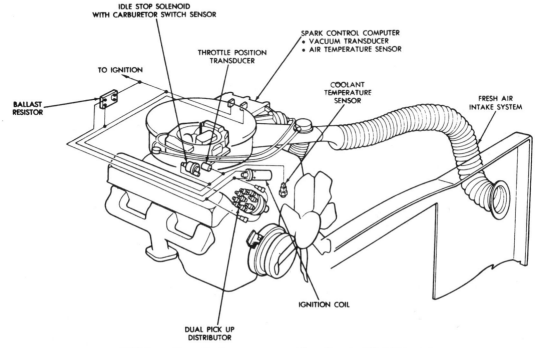

1976 Lean Burn System component locations—400 4 bbl engine

The Run Pick-up Sensor, also located in the distributor, provides timing data to the computer once the engine is running. It also monitors engine speed and helps the computer decide when the piston is reaching the top of its compression stroke.

Coolant Temperature Sensor, located on the water pump housing, informs the computer when the coolant temperature is below 150°.

Air Temperature Sensor, inside the computer itself, monitors the temperature of the air coming into the air cleaner.

Throttle Position Transducer, located on the carburetor, monitors the position and rate of change of the throttle plates. When the throttle plates start to open and as they continue to open toward full throttle, more and more spark advance is called for by the computer. If the throttle plates are opened quickly even more spark advance is given for about one second. The amount of maximum advance is determined by the temperature of the air coming into the air cleaner. Less advance under acceleration will be given if the air entering the air cleaner is hot, while more advance will be given if the air is cold.

Carburetor Switch Sensor, located on the end of the idle stop solenoid, tells the computer if the engine is at idle or off-idle.

Vacuum Transducer, located on the computer, monitors the amount of intake manifold vacuum present; the more vacuum, the more spark advance to the distributor. In order to obtain this spark advance in the distributor, the carburetor switch sensor has to remain open for a specified amount of time during which the advance will slowly build up to the amount indicated as necessary by the vacuum transducer. If the carburetor switch should close during that time, the advance to the distributor will be cancelled. From here the computer will start with an advance countdown. If the carburetor switch is reopened within a certain amount of time, the advance will continue from a point where the computer decides it should. If the switch is reopened after the computer has counted down to "no advance," the vacuum advance process must start over again.

Operation

When you turn the ignition key on, the start pick-up sends its signal to the computer which relays back information for more spark advance during cranking. As soon as the engine starts, the run pick-up takes over and receives more advance for about one minute. This advance is slowly eliminated during the one minute warm-up period. While the engine is cold (coolant temperature below 150° as monitored by the coolant temperature sensor), no more advance will be given to the distributor until the engine reaches normal operating temperature, at which time normal operation of the system will begin.

In normal operation, the basic timing information is relayed by the run pick-up to the computer along with input signals from all the other sensors. From this data the computer determines the maximum allowable advance or retard to be sent to the distributor for any situation.

If either the run pick-up or the computer should fail, the back-up system of the start pick-up takes over. This supplies a fixed timing signal to the distributor which allows the car to be driven until it can be repaired. In this mode, very poor fuel economy and performance will be experienced. If the start pick-up or the ignition control module section of the computer should fail, the engine will not start or run.

System Tests, Adjustments, and Service

The following sections give tests, adjustments, and/or replacement of some of the emission control system components used on your Plymouth which may be easily owner-serviced.

It is important to remember that some things listed above cannot be serviced, as they are part of the engine design; and in other cases, because of system complexity, service is best left to qualified personnel.

POSITIVE CRANKCASE VENTILATION SYSTEM (PCV)

Testing

1. With the engine running at curb idle, unplug the PCV valve from the

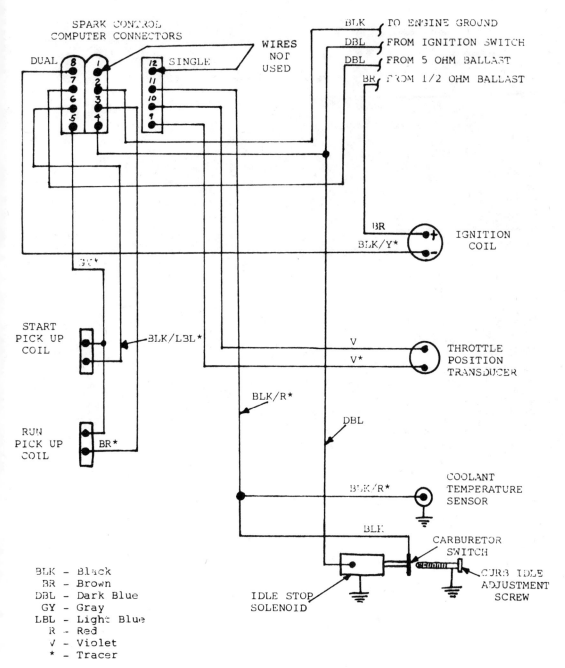

1976 Lean Burn System schematic

rocker cover and check for vacuum at the open end of the valve, with your finger or a vacuum gauge. If the vacuum is weak, renew the valve or hose, whichever is needed.

2. Plug the PCV valve back into the rocker cover.

3. Remove the breather cap from the rocker cover. The cap to remove is the large one with one or more hoses connected to it. If there is a separate oil filler cap, it must stay in the closed position on the rocker cover.

4. On 1970–71 6 cylinder engines, pinch off, or clamp shut, the hose between the carburetor float bowl and the fuel pump casting.

5. With the engine idling, put a sheet

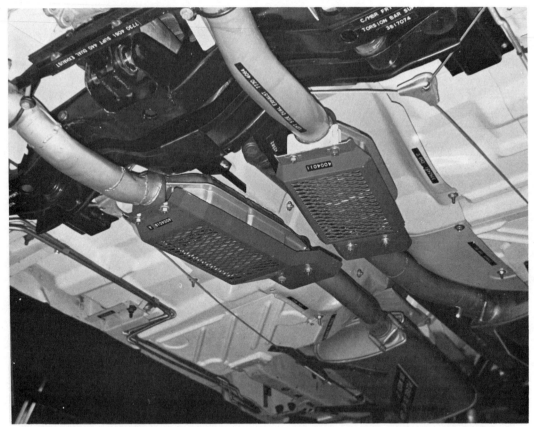

Catalytic converter installation on 1975–76 models with dual exhaust—converters are in front of regular mufflers

of paper or a vacuum gauge over the breather opening in the rocker cover to check crankcase vacuum. It may take as long as a minute for vacuum to build up. If there is not enough vacuum to suck the paper against the hole, check for air leaks at the engine gasket or the dipstick.

6. Blow through the fresh air hose and the breather cap to be sure that they are clear, then install the breather cap on the rocker cover. Release the pinch or clamp on the 6 cylinder hose.

Servicing

Do not attempt to clean the PCV valve. If it is defective or if the mileage interval specified under "Routine Maintenance" in chapter 1 has been reached, replace the valve with a new one as outlined below. Remember that not all PCV valves are interchangeable, so be sure to specify the engine size and model year when purchasing a new valve.

To replace the valve:

1. Remove the PCV valve from its grommet on the valve cover.

2. Pull the valve out of the hose, after loosening the hose clamp.

3. Install the new valve in the reverse order of removal.

4. Clean the breather cap filter, as outlined under "Routine Maintenance" in Chapter 1.

AIR INJECTION SYSTEM

Component Testing

CAUTION: *Do not hammer, pry, or bend the air pump housing while tightening the drive belt or testing the pump.*

BELT TENSION AND AIR LEAKS

1. Before proceeding with the tests, check the pump drive belt tension to see that it is not loose.

Pull the PCV valve out of the rocker cover to test it

2. Turn the pump by hand. If it has seized, the belt will slip, making a noise. Disregard any chirping, squealing, or rolling sounds from inside the pump; these are normal when it is turned by hand.

3. Check the hoses and connections for leaks. Hissing or a blast of air is indicative of a leak. Soapy water, applied lightly around the area in question, is a good method for detecting leaks.

PUMP AIR OUTPUT

1. Disconnect the air supply hose at the antibackfire valve.

2. Connect a pressure gauge, using a suitable adapter, to the air supply hose. NOTE: *If there are two hoses, plug the second one.*

3. With the engine at normal operating temperature, increase the idle speed and watch the vacuum gauge.

4. The airflow from the pump should be steady and fall between 2 and 6 psi. If it is unsteady or falls below this, the pump is defective and must be replaced.

PUMP NOISE DIAGNOSIS

The air pump is normally noisy; as engine speed increases, the noise of the pump will rise in pitch. The rolling sound the pump bearings make is normal. But if this sound becomes objectionable at certain speeds, the pump is defective and will have to be replaced.

A continual hissing sound from the air pump pressure relief valve at idle, indicates a defective valve. Replace the relief valve.

If the pump rear bearing fails, a continual knocking sound will be heard. Since the rear bearing is not separately replaceable, the pump will have to be replaced as an assembly.

RELIEF VALVE

NOTE: *If the relief valve does not leak at idle it is probably not defective. Do not increase engine speed in order to make the relief valve open.*
1. Disconnect the output hose from the pump. If there are two output hoses, disconnect both.

2. With the engine idling, block the pump hose(s) with your finger in order to make the relief valve open. You should hear air escaping from the valve.

3. Remove your finger, the valve should close and be quiet.

NOTE: *If the pump has been replaced there may be relief valves in the pump and diverter valve, as well. Whenever you find two relief valves, you should check both of them for leaks; but only one of them will open to relieve pressure, usually the one on the diverter valve. The second valve in the pump may be ignored as long as it does not leak.*

DIVERTER (BY-PASS) VALVE

1. Detach the hose, which runs from the diverter valve to the check valve, at the diverter valve hose connection.

2. Connect a tachometer to the engine. With the engine running at normal idle speed, check to see that air is flowing from the valve hose connection.

3. Speed the engine up, so that it is running at 1,500–2,000 rpm. Allow the throttle to snap shut. The flow of air

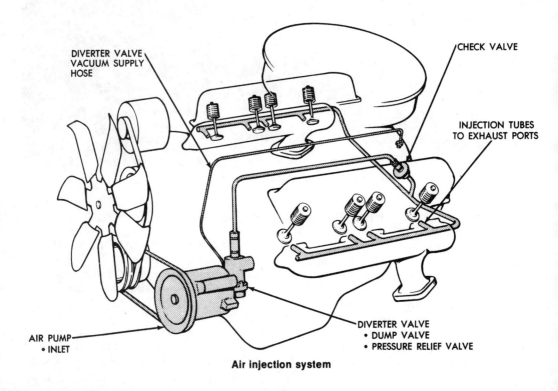

DIVERTER VALVE
VACUUM SUPPLY
HOSE

CHECK VALVE

INJECTION TUBES
TO EXHAUST PORTS

AIR PUMP
• INLET

DIVERTER VALVE
• DUMP VALVE
• PRESSURE RELIEF VALVE

Air injection system

from the diverter valve at the check valve hose connection should stop momentarily and air should then flow from the exhaust port on the valve body or the silencer assembly.

4. Repeat Step 3 several times. If the flow of air is not diverted into the atmosphere from the valve exhaust port or if it fails to stop flowing from the hose connection, check the vacuum lines and connections. If these are tight, the valve is defective and requires replacement.

5. A leaking diaphragm will cause the air to flow out both the hose connection and the exhaust port at the same time. If this happens, replace the valve.

CHECK VALVE

1. Before starting the test, check all of the hoses and connections for leaks.

2. Detach the air supply hose(s) from the check valve(s).

3. Insert a suitable probe into the check valve and depress the plate. Release it; the plate should return to its original position against the valve seat. If binding is evident, replace the valve.

4. Repeat Step 3 if two valves are used.

5. With the engine running at normal operating temperature gradually increase its speed to 1,500 rpm. Check for exhaust gas leakage. If any is present, replace the valve assembly.

NOTE: *Vibration and flutter of the check valve at idle speed is a normal condition and does not mean that the valve should be replaced.*

Removal and Installation

AIR PUMP

CAUTION: *Never place the pump in a vise or attempt to dismantle it. The pump has no internal parts that are replaceable and it is serviced as a unit. Never pry or hammer on the pump housing.*

1. Loosen the bolts on the pump pulley.

2. Loosen the air pump attachment bracket.

3. Detach the air supply hoses at the pump.

4. Remove the drivebelt and pulley from the hub.

5. Unfasten the bolts on the bracket and withdraw the pump.

Installation is performed in the following order:

1. Place the pump on its mounting bracket and install, but do not tighten, the attachment bolts.

2. With the rotor shaft used as a center, fit the pulley into the hub and install the drivebelt over the pulley.

3. Tighten the pulley attachment bolts to 95 in. lbs, using care not to snap them off.

4. Adjust the pump until the belt is secure. Tighten the mounting bolts and the adjusting screw to 30 ft lbs maximum; do not overtighten.

5. Attach the hoses and clamps.

PUMP-MOUNTED RELIEF VALVE

NOTE: *The pumps used on Plymouths do not have relief valves mounted on their housings. Instead, the relief valves are part of the diverter valve assemblies.*

CENTRIFUGAL FILTER FAN

NOTE: *Never attempt to clean the filter fan. It is impossible to remove the fan without destroying it.*

1. Remove the air pump from the car, as detailed above.

2. Gently pry the outer disc off and pull off the remaining portion. Be careful that no fragments from the fan enter the pump air intake.

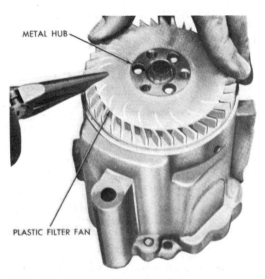

METAL HUB

PLASTIC FILTER FAN

Removing the centrifugal filter fan

3. Install a new filter fan, pulling it into place with the pump pulley and attaching bolts.

4. Alternately torque the bolts so that the fan is drawn down *evenly*. Be sure that the outer edge of the fan fits into the pump housing.

CAUTION: *Never hammer or press the fan into place; damage to it and the pump will result.*

5. Install the pump on the car.

NOTE: *For the first 20–30 miles of operation, the fan may squeal until its lip has worn in. This is normal and does not indicate a damaged pump.*

EXHAUST TUBE

1. Remove the exhaust tube by grasping *it* (never the pump body) in a vise or a pair of pliers. Pull the tube out with a gentle twisting motion.

2. Install the new exhaust tube by tapping it into the hole with a hammer and a wooden block. Be careful not to damage its end.

3. Tap it until ⅞ in. of the tube remains above the pump cover.

CAUTION: *Do not clamp the pump in a vise while installing the exhaust tube.*

DIVERTER (BY-PASS) VALVE

1. Disconnect the hoses from the valve.

2. Remove the screws that attach the valve bracket to the engine or the back of the pump. Withdraw the valve and bracket assembly.

3. Installation is the reverse of removal.

THERMOSTATICALLY CONTROLLED AIR CLEANER

Testing

1. Remove the air cleaner from the carburetor and allow it to cool to 90° F. Connect a vacuum source to the sensor as well as a vacuum gauge.

2. Apply 20 in. Hg to the sensor, the door should be in the "heat on" (up) position. If it remains in the "off" position, test the vacuum motor.

3. Connect the motor to a vacuum source. In addition to the vacuum gauge, a hose clamp and a bleed valve are necessary. Connect them in the following order:

a. Vacuum source

b. Hose clamp (or shut-off valve)

c. Bleed valve

Air Injection System Diagnosis Chart

Problem	Cause	Cure
1. Noisy drive belt	1a Loose belt 1b Seized pump	1a Tighten belt 1b Replace
2. Noisy pump	2a Leaking hose 2b Loose hose 2c Hose contacting other parts 2d Diverter or check valve failure 2e Pump mounting loose 2g Defective pump	2a Trace and fix leak 2b Tighten hose clamp 2c Reposition hose 2d Replace 2e Tighten securing bolts 2g Replace
3. No air supply	3a Loose belt 3b Leak in hose or at fitting 3c Defective anti-backfire valve 3d Defective check valve 3e Defective pump	3a Tighten belt 3b Trace and fix leak 3c Replace 3d Replace 3e Replace
4. Exhaust backfire	4a Vacuum or air leaks 4b Defective anti-backfire valve 4c Sticking choke 4d Choke setting rich	4a Trace and fix leak 4b Replace 4c Service choke 4d Adjust choke

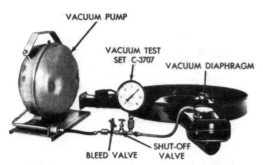

Test hook-up for the air cleaner door vacuum motor

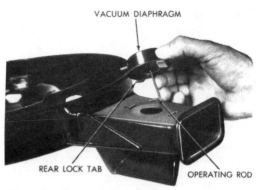

Removing the vacuum motor

    d. Vacuum gauge
    e. Vacuum motor
    4. Apply 20 in. Hg vacuum to the motor. Use the hose clamp to block the line, so that the motor will retain the vacuum. The door operating motor should retain this amount of vacuum for five minutes. Release the hose clamp.

NOTE: *If the vacuum cannot be built up to the specified amount, the diaphragm has a leak and the valve will require replacement.*

    5. By slowly closing the bleed valve, check the operation of the door. The door should start to raise at not less than 5 in. Hg and should be fully raised at no more than 9 in. Hg.
    6. If the vacuum motor fails any of the tests in Steps 3–5, it is defective. Replace it with a new unit.
    7. If the door works properly but fails to pass Step 2, the sensor is at fault and should be replaced.

NOTE: *If the engine has a dual snorkle air cleaner, check the right side as in steps 3–5, above. However, there is no temperature sensor on the right-side door.*

Replacement

VACUUM MOTOR

    1. Remove the air cleaner assembly from the carburetor.
    2. Disconnect the vacuum hose from the vacuum motor.
    3. To disengage its lock, tip the motor slightly forward and turn it counterclockwise.
    4. Once the motor is clear of the snorkle, slide it to one side, so that its linkage is disengaged from the air door.
    5. Installation of a new vacuum motor is performed in the reverse order of removal of the old one.

TEMPERATURE SENSOR

    1. Remove the air cleaner assembly from the carburetor.
    2. Lift the top cover off the air cleaner and disconnect the vacuum lines from the temperature sensor.
    3. Unfasten and discard the clips securing the sensor.
    4. Remove the old sensor, complete with its gasket.
    5. Installation, using the new gasket and clips supplied, of a new sensor is the reverse of the removal of the old one.

TIMING RETARD SOLENOID TEST

A timing retard solenoid is uded on some Plymouths up to and including the 1971 model year.
    1. Connect a timing light to the engine and check the timing.
    2. Detach the solenoid ground lead near its carburetor end. Timing should advance at least 5° and there should be an increase in engine speed.
    3. Reconnect the ground lead. Timing should be retarded to the original position noted and the engine should slow down. Repeat the test several times.
    4. If the timing does not behave in the manner indicated, the solenoid is defective and must be replaced; it cannot be adjusted or repaired.

TIMING ADVANCE SOLENOID TEST

NOTE: *A timing advance solenoid is used on Plymouths, starting in 1972. It should not be confused with the re-*

tard solenoid, above. It is used only in the 400 cu. in. 4-bbl engine in 1973–1974.

1. Attach a tachometer to the engine.

2. Detach the vacuum advance line from the distributor advance unit.

3. Run the engine at normal idle and check engine rpm with the tachometer.

4. Detach the solenoid lead wire at its connection, which is about 6 in. away from it. Run a jumper wire from the battery to the disconnected lead from the solenoid.

5. When the two leads are touched,

DISCONNECT HERE

Disconnect the timing advance solenoid where indicated

the engine speed should increase by 50 rpm or, if a timing light is used, the timing should advance 7–8°.

NOTE: *Do not touch the jumper wire to the solenoid lead for more than 30 seconds, or the solenoid will overheat.*

6. If the engine speed does not increase or the timing advance, the solenoid is defective and must be replaced. Remember to reconnect it when through testing.

DECELERATION VALVE

Testing

NOTE: *Timing, idle speed and air/fuel mixture should be at proper specifications before starting this test.*

1. Connect a vacuum gauge to the distributor vacuum advance line, by using a T-connection which has about the same inside diameter as the line. Do not clamp the line shut.

2. If the carburetor is equipped with a dashpot, tape its plunger down so that it cannot touch the throttle lever at idle.

3. Speed the engine up to about 2000 rpm and retain this speed for about five seconds.

4. Release the throttle, allowing the engine to return to normal idle.

5. The vacuum reading should rise to about 15–16. in. Hg and stay there for one second. It should take about three seconds for the vacuum to return to its normal 6 in. Hg reading.

6. If the valve does not retain its high reading for about one second or if it takes over three seconds for the reading to return to normal, the valve should be adjusted as outlined below.

To check for a leaking valve diaphragm, proceed as follows:

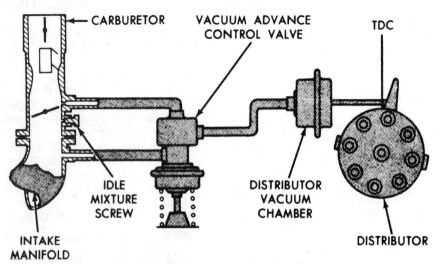

Deceleration valve vacuum circuit

1. Remove the vacuum gauge and connect it to the manifold vacuum line with a T-connection.

2. Clamp the valve-to-distributor vacuum line and, with the engine at normal idle speed, check the vacuum reading.

3. Clamp the line shut between the deceleration valve and the T-connection. Check the vacuum gauge reading again.

4. If the second reading is higher than the first, the valve diaphragm is leaking and the valve should be replaced.

Adjusting

If the deceleration valve test indicated a need for adjustments, proceed as follows:

1. Remove the cover to gain access to the adjusting screw.

2. If an *increase* in valve opening time is desired, turn the adjusting screw counterclockwise.

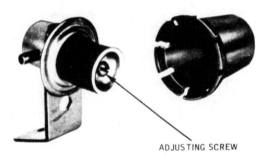

ADJUSTING SCREW

Deceleration valve adjusting screw

3. If a *decrease* in time is desired, turn the adjusting screw clockwise.

NOTE: *Each complete turn of the adjusting screw equals ½ in. Hg. Thus, if the vacuum reading at the end of three seconds is 7½ in. Hg, it will take three turns of the screw to return it to the proper 6 in. Hg reading.*

4. After finishing the adjustments, retest the valve, as outlined above. If the valve cannot be adjusted to proper specifications, it is defective and must be replaced.

ORIFICE SPARK ADVANCE CONTROL (OSAC) VALVE TESTS

NOTE: *Air temperature around the car must be above 68° F for this test because the OSAC valve has a temperature sensor in it. Valves produced from around 1 March 1973 do not* have an ambient temperature sensor built into them. They may be identified by their white gasket (old OSAC valves had a black one) and a pasted-on label with the part number (3755499) on it. 1974–76 models have the OSAC valve located in the air cleaner, rather than on the firewall and are also affected by ambient temperature (See description above).

1. Check the vacuum hoses and connections for any signs of leaks or plugging.

2. Detach the vacuum line which runs from the distributor to the OSAC valve at the distributor end. Connect a vacuum gauge to this line.

3. Connect a tachometer to the engine. Rapidly open the throttle and then stabilize the engine speed at 2000

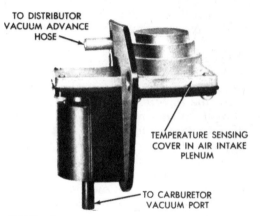

TO DISTRIBUTOR VACUUM ADVANCE HOSE

TEMPERATURE SENSING COVER IN AIR INTAKE PLENUM

TO CARBURETOR VACUUM PORT

OSAC valve

rpm in neutral. When the throttle is rapidly opened the vacuum gauge reading should drop to zero.

4. With the engine speed at a steady 2000 rpm, it should take about 15 seconds for the vacuum level to rise and then stabilize.

NOTE *The length of time may vary slightly with different engines; 15 seconds is an approximate figure.*

5. If the vacuum level rises immediately, the valve is defective and must be replaced.

6. If there is no increase in vacuum at all, disconnect the hose which runs from the carburetor to the OSAC valve at the valve and connect a vacuum gauge to this hose. Speed the engine up to 2000 rpm.

7. If there is no vacuum reading on

the gauge, check for a clogged carburetor port, filters or hoses.

8. If there is a vacuum reading, the valve is defective and must be replaced.

9. Reconnect the vacuum hoses, after disconnecting the vacuum gauge. Disconnect the tachometer.

IDLE STOP SOLENOID TESTS

CAUTION: *Chrysler Corp. cars that are equipped with catalytic converters have an additional solenoid; this is NOT an idle stop solenoid, and no attempt to adjust the idle speed with it should be made. See "Catalytic Converter—1975–76" for its operation.*

1. Turn the ignition key on and open the throttle. The solenoid plunger should extend (solenoid energize).

2. Turn the ignition off. The plunger should retract, allowing the throttle to close.

NOTE: *With the idle stop solenoid deenergized, the carburetor idle speed adjusting screw must make contact with the throttle shaft to prevent the throttle plates from jamming in the throttle bore when the engine is turned off.*

3. If the solenoid is functioning properly and the engine is still dieseling, check for one of the following:

    a. High idle or engine shut off speed;

    b. Engine timing not set to specification;

    c. Binding throttle linkage;

    d. Too low an octane fuel being used.

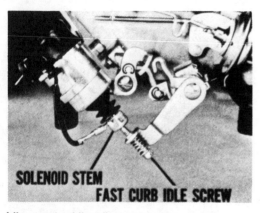

SOLENOID STEM
FAST CURB IDLE SCREW

Idle stop (antidieseling) solenoid

Correct any of these problems, as necessary.

4. If the solenoid fails to function as outlined in Steps 1–2, disconnect the solenoid leads; the solenoid should deenergize. If it does not, it is jammed and must be replaced.

5. Connect the solenoid to a 12 V power source and to ground. Open the throttle so that the plunger can extend. If it does not, the solenoid is defective.

6. If the solenoid is functioning correctly and no other source of trouble can be found, the fault probably lies in the wiring between the solenoid and the ignition switch or in the ignition switch itself. Remember to reconnect the solenoid when finished testing.

ELECTRICALLY ASSISTED CHOKE TESTS

CAUTION: *Do not immerse the choke heating element in any type of liquid, especially solvent, for any reason.*

NOTE: *A short circuit in the choke wiring or in the heater will show up as a short in the ignition system.*

1. Disconnect the electrical leads from the choke control switch before starting the engine.

2. Connect a test light between the smaller of the two terminals on the choke control switch and a ground.

3. Start the engine and run it until it reaches normal operating temperature.

4. Apply power from a 12 V source to the terminal marked "BAT" on the choke control switch.

5. The test light should light for at least a few seconds or for as long as five minutes. If the light does not come on at all or if it stays on longer than five minutes, replace the switch.

6. Disconnect the test light and reconnect the electrical leads to the choke switch, if it is functioning properly.

7. Detach the lead from the choke switch which runs to the choke heating element.

8. Connect the lead from an ohmmeter to the crimped section at the choke end of the wire, which was removed in Step 7.

CAUTION: *Do not connect the ohmmeter to the metallic heater housing.*

9. Ground the other ohmmeter test lead to the engine manifold.

10. The meter should indicate a resistance of 4–6 ohms.

11. If the reading is not within specifications, or if it indicates an opened (zero resistance) or a shorted (infinite resistance) heater coil, replace the heater assembly.

NOTE: *The electrically assisted choke does not change any carburetor service procedures. If any parts of the electrically assisted choke are defective, they must be replaced. Adjustment is not possible.*

1975–76 COOLANT CONTROL IDLE ENRICHMENT (CCIE) SYSTEM

California System Test

NOTE: *The engine coolant temperature must be below 80° F for this test.*

1. Disconnect the vacuum line, which runs from the two-nozzle coolant temperature operated vacuum valve to the CCIE valve on the carburetor, at the air bleed (on the vacuum valve side).

2. Connect a vacuum gauge to this line and start the engine.

3. The vacuum gauge should register manifold vacuum. If it does not, check the lines and fittings to see if they are pinched or plugged.

4. If everything is in good condition, replace the coolant temperature operated vacuum valve.

5. If the valve is OK, then run the engine until the coolant temperature goes above 90° F. Above this temperature, the vacuum gauge should read zero; if it does not, replace the vacuum valve.

6. Check the air bleed and the vacuum line which runs from it to the carburetor. Replace either if it is plugged.

7. Remove the vacuum gauge and connect the vacuum lines.

Federal System Test

NOTE: *The engine coolant temperature must be below 150° F for this test.*

1. Disconnect the vacuum line which runs from the two-nozzle coolant temperature operated vacuum valve to the CCIE valve on the carburetor, at the air bleed (on the vacuum valve side).

2. Connect a vacuum gauge to the disconnected line.

3. Start the engine, the gauge should

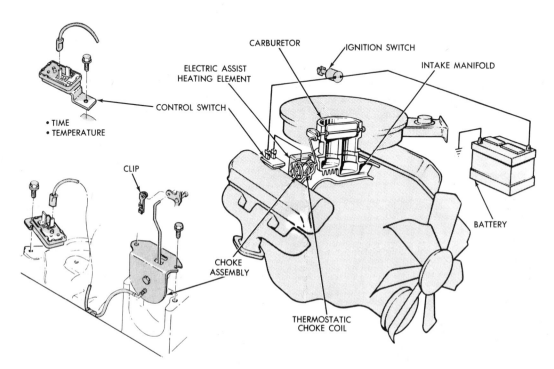

Electrically-assisted choke system

CCIE valve on 1975–76 carburetor—note its vacuum line air bleed

register manifold vacuum for about 35 seconds, and then drop to zero.

4. If there is no vacuum, stop the engine. Check the vacuum lines and fittings. If these are in good shape, check the EGR delay timer and solenoid as detailed in the appropriate section.

5. If the delay timer, solenoid, fittings, and lines check out OK, then replace the two-nozzle coolant temperature operated vacuum valve.

6. If there was vacuum present long after 35 seconds in Step 3, then check the EGR delay timer and solenoid.

7. Start the engine and allow the coolant to warm-up to above 150° F, then stop the engine.

8. Start the engine again; there should be no vacuum reading on the gauge. If there is, replace the coolant temperature operated vacuum valve.

9. Check the air bleed and the vacuum line from it to the carburetor. Replace either if it is plugged.

10. Remove the vacuum gauge and connect the vacuum lines.

EVAPORATIVE EMISSIONS CONTROL (EEC) SYSTEM

About the only regularly scheduled service required for the EEC system, is replacement of the filter used on the charcoal canister system. For a description of this operation, including specified service intervals, see "Routine Maintenance" in Chapter 1.

Testing

There are several things to check for if a malfunction of the evaporative emission control system is suspected.

1. Leaks may be traced by using an infrared hydrocarbon tester. Run the test probe along the lines and connections. The meter will indicate the presence of a leak by a high hydrocarbon (HC) reading. This method is much more accurate than a visual inspection which would in-

dicate only the presence of a leak large enough to pass liquid.

2. Leaks may be caused by any of the following, so always check these areas when looking for them:

    a. Defective or worn lines;

    b. Disconnected or pinched lines;

    c. Improperly routed lines;

    d. A defective filler cap.

NOTE: *If it becomes necessary to replace any of the lines used in the evaporative emission control system, use only those hoses which are fuel resistant or are marked "EVAP."*

3. If the fuel tank has collapsed, it may be the fault of clogged or pinched vent lines, a defective check valve or a plugged or incorrect fuel filler cap.

4. To test the filler cap, clean it and place it against the mouth. Blow into the relief valve housing. If the cap passes pressure with light blowing or if it fails to release with hard blowing, it is defective and must be replaced.

NOTE: *Replace the cap with one marked "pressure/vacuum" only. An incorrect cap will render the system inoperative or damage its compo-*

Removal and Installation

Removal and installation of the various evaporative emission control system components consists of unfastening hoses, loosening securing screws, and removing the part which is to be replaced from its mounting bracket. Installation is the reverse of removal.

NOTE: *When replacing any EEC system hoses, always use hoses that are fuel-resistant or are marked "EVAP."*

1972–73 EGR FLOOR JET SERVICE

All six-cylinder engines have one floor jet, while all V8s have two.

1. Turn the engine off. Remove the air cleaner assembly from the carburetor.

2. Hold the choke and throttle valves open. Shine a flashlight through the carburetor to inspect the floor jet(s). The jet(s) is/are in satisfactory condition if the passage shows an open path to the orifice.

3. If the jet(s) is/are clogged, completely remove the carburetor. Withdraw the jet and clean it.

CAUTION: *Use care when handling the jets. They have very thin walls and are, therefore, easily damaged. Because they are made out of stainless steel, they are not magnetic and cannot be retrieved readily if dropped into the manifold.*

4. Install the jet(s) and tighten to 25 ft lbs. Install the carburetor and attach the air cleaner.

NOTE: *"Shorting out" cylinders on engines equipped with floor jets is not*

EGR floor jets

a reliable test procedure. The un-burned mixture is circulated to the other cylinders, causing the engine speed to fluctuate. Because of this, false test results may be obtained.

1973 PROPORTIONAL EGR SYSTEM TESTS

NOTE: *Air temperature should be above 68° F for this test.*

1. Check all of the vacuum hoses which run between the carburetor, intake manifold, EGR valve, and the vacuum amplifier (if so equipped). Replace the hoses and tighten the connections, as required.

2. Allow the engine to warm up. Connect a tachometer to it. Start with the engine idling in neutral and rapidly increase the engine speed to 2,000 rpm.

3. If the EGR valve stem moves (watch the groove on the stem), the valve and the rest of the system are functioning properly. If the stem does not move, proceed with the rest of the EGR system tests.

4. Disconnect the vacuum supply hose from the EGR valve. Apply a vacuum of at least 10 in. Hg to the valve with the engine warmed-up and idling and the transmission in neutral.

NOTE: *A source of more than adequate vacuum is the intake manifold vacuum connection. Run a hose from the EGR valve directly to the connection.*

5. When vacuum is applied to the EGR valve, the engine speed should drop at least 150 rpm. In some cases the engine may even stall. If the engine does not slow down and the EGR valve does not operate, the valve is defective or dirty. Replace it or remove the deposits from it.

NOTE: *Always replace the EGR valve gasket with a new one when the valve is removed for service, even if the valve itself is not replaced.*

6. If the EGR valve is functioning properly, reconnect its vacuum line and test the temperature control valve.

7. Disconnect the vacuum hose which runs to the temperature control valve and plug it. Repeat steps 2–3. If the EGR valve now functions, the temperature control valve is defective and must be replaced.

8. If everything else is functioning properly, the EGR system does not work and the engine is equipped with a vacuum amplifier (see the chart below), the amplifier is at fault. Replace it and repeat the system test.

NOTE: *Before replacing the amplifier, check the vacuum port in the carburetor. If it is clogged, clean it with solvent; do not use a drill.*

Engine	Vacuum amplifier used on:
225—6	All
318—V8	California only
360—V8	All
440—V8	All non-high-performance engines

1973½–76 PROPORTIONAL EGR SYSTEM TESTS

NOTE: *This system is used starting with cars made on or after 15 March 1973. It replaces the system tested in the above section.*

1. Perform Steps 1–6 of the 1973 proportional EGR system tests, detailed in the above section.

2. Test the EGR system coolant temperature operated control valve for leaks. The valve is located on either the right or left side of the radiator top tank.

3. Disconnect the vacuum hose from the EGR coolant temperature operated control valve, then connect a vacuum source and gauge to the valve fitting, in place of the hose.

4. Apply 10 in. Hg of vacuum to the valve. If the valve loses more than 1 in. Hg in one minute, the valve is defective and must be replaced.

5. Proceed with Step 8 in the above section.

1974–76 EGR DELAY SYSTEM TESTS

NOTE: *Not all engines use an EGR delay timer; some 1975–76 engines also used the EGR delay timer to control the coolant control idle enrichment (CCIE) system as well.*

1. Unfasten the distributor-to-coil lead.

2. Disconnect the vacuum line which runs from the delay solenoid to the vacuum amplifier at the amplifier end.

3. Turn the car's ignition switch to

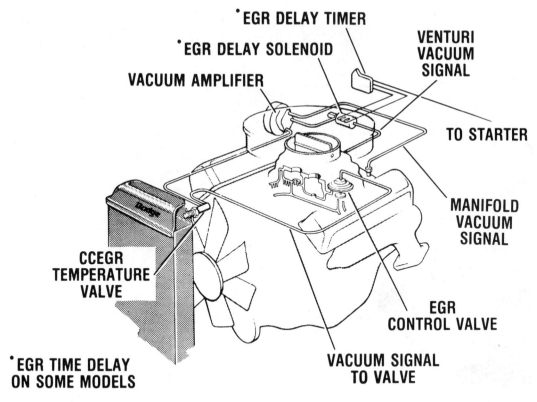

EGR DELAY TIMER

EGR DELAY SOLENOID

VENTURI VACUUM SIGNAL

VACUUM AMPLIFIER

TO STARTER

MANIFOLD VACUUM SIGNAL

CCEGR TEMPERATURE VALVE

EGR CONTROL VALVE

EGR TIME DELAY ON SOME MODELS

VACUUM SIGNAL TO VALVE

1973½–76 proportional EGR system

"START" and then release it, so that it returns to "RUN."

4. Suck on the end of the disconnected hose; the hose should be blocked.

5. After about 35 seconds from the time that the ignition switch was turned to "START," the solenoid should open, allowing air to flow through the line that you are sucking on.

6. If the system isn't working, disconnect the solenoid and connect it directly to a 12-volt power source, making and breaking the circuit several times. If the solenoid works, replace the delay timer.

7. If the solenoid doesn't work, replace the solenoid.

8. Reconnect the vacuum lines and the coil after completing the test.

1975–76 EGR REMINDER LIGHT

NOTE: *This light is designed to remind you that regularly scheduled service is due; it does not mean that the EGR system is not working properly.*

1. After checking the EGR system for proper operation, slide the rubber boot on the EGR reminder odometer up, out of the way.

2. Reset the odometer with a small screwdriver.

3. Slide the boot back down over the odometer. The light will come on again when the next 15,000 mile check-up is due.

1975 CATALYST PROTECTION SYSTEM TESTING

1. Start the engine and allow it to warm up.

2. Connect a tachometer to the engine.

3. Increase the engine speed to 2500 rpm; the solenoid plunger should extend.

4. Release the throttle, once the engine speed has dropped below 2000 rpm; the plunger should retract.

5. If the solenoid doesn't work properly, disconnect it from the speed switch and connect it to a 12 volt power source. If the plunger extends, the solenoid is working and the speed switch is defective. Replace the speed switch.

The large solenoid in the lower left-hand corner is part of the catalyst protection system used only on 1975 models—do not adjust the idle speed with it.

6. If the plunger does not extend, the solenoid is defective and must be replaced.

7. With the solenoid connected to the 12 volt power source (energized) so that its plungers is out, adjust the engine speed to 1,500 rpm.

CAUTION: *Do not adjust the curb idle speed with the catalyst protection system solenoid. This is not an anti-dieseling solenoid. If the curb idle speed is adjusted with this solenoid, it will be far above 2,000 rpm once the solenoid has been energized, creating a hazardous driving situation. The only way to stop an engine which has been adjusted in this manner, is to shut it off and not restart it until the solenoid has been correctly adjusted.*

8. Reconnect the solenoid to the speed switch and disconnect the tachometer, once the proper setting has been obtained.

REPLACING THE CATALYST

The catalyst used on Plymouths is the monolithic (one-piece) type which cannot be removed from the converter for replacement.

If the catalyst fails, it will be necessary to replace the entire converter assembly with a new one. To do so, proceed as follows:

CAUTION: *Allow the converter assembly to cool completely before attempting to service it; catalyst temperatures can reach 1500°–1600° F.*

1. If a grass shield is used, remove the bolts which secure it and lower the shield from underneath the vehicle.

2. Unbolt the converter assembly at the mounting flanges, just as you would a normal exhaust pipe from the manifold.

NOTE: *Support the exhaust pipe while the converter is removed.*

3. Replace the old converter with a

Remove the four bolts securing the converter grass shields

new unit which is the exact same size, part number, and type (or its *exact* equivalent) as was originally installed.

4. Remove the plastic plugs from the ends of the new converter (if used) and install it in the reverse order of removal, being sure to use all required gaskets to ensure a leak-free fit.

5. Install the grass shields.

LEAN BURN SYSTEM TESTING AND SERVICE

Some of the procedures in this section refer to an adjustable timing light. This is also known as a spark advance tester, i.e., a device which will measure how much spark advance is present going from one point, a base figure, to another. Since precise timing is very important to the Lean Burn System, do not attempt to perform any engine tests calling for an adjustable timing light without one. In places where a regular timing light can be used, it will be noted in the text.

Troubleshooting

1. Remove the coil wire and hold it about ¼ in. away from an engine ground, then have someone crank the engine while you check for spark.

2. If you have a good spark, slowly move the coil wire away from the engine and check for arcing at the coil while cranking.

3. If you have good spark and it is not arcing at the coil, check the rest of the parts of the ignition system, if they are alright, the problem is not in the ignition system. Check the "Troubleshooting" section following Chapter 2.

ENGINE NOT RUNNING—WILL NOT START

1. Check the battery specific gravity; it must be at least 1.220 to deliver the necessary voltage to fire the plugs.

2. Remove the terminal connector from the coolant switch and put a piece of paper or plastic between the curb idle adjusting screw and the carburetor switch. This is unnecessary if the screw and switch are not touching.

3. Connect the negative lead of a voltmeter to a good engine ground, turn the ignition switch to the "Run" position and measure the voltage at the carburetor switch terminal. If you receive a reading of more than five volts, go on to Step 7; if not, proceed to the next step.

4. Turn the ignition switch "Off" and

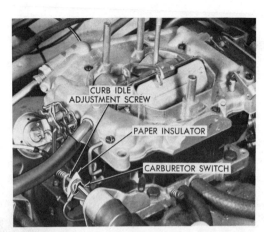

Preparing for power check

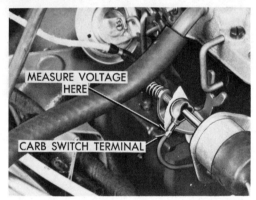

Power check

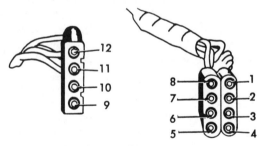

Connector terminal numbering

disconnect the double terminal connector from the bottom of the Spark Control Computer. Turn the ignition switch back to the "Run" position and measure the voltage at terminal No. 4; if the voltage is not within 1 volt of the voltage you received in Step 3, check the wiring between terminal No. 3 and the ignition switch. If the voltage is correct, go on to the next step.

5. Turn the ignition switch "Off" and disconnect the single connector from the bottom of the Spark Control Computer. Using an ohmmeter, check for continuity between terminal No. 11 and the carburetor switch terminal. There should be continuity present, if not, check the wiring.

6. Check for continuity between terminal No. 2 (double connector) and ground. If there is continuity, replace the Spark Control Computer; if not, check the wiring. If the engine will not start, proceed to the next step.

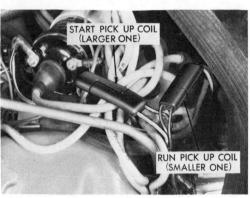

Pick-up coil connector identification

7. Turn the ignition switch to the "Run" position and check for voltage at terminals Nos. 7 and 8 of the double connector. If you received voltage within 1 volt of that recorded in Step 3, proceed to the next step. If you did not on terminal No. 7, check the wiring between it and the ignition switch and check the 5 ohm side of the ballast resistor. If you did not on terminal No. 8, check the wiring, and the primary windings of the coil and the ½ ohm side of the ballast resistor.

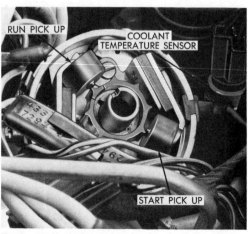

Engine sensors

8. Turn the ignition switch "Off" and with an ohmmeter, measure resistance between terminals Nos. 5 and 6 of the dual connector. If you do not receive a reading of 150–900 ohms, disconnect the start pick-up leads at the distributor and measure the resistance going into the distributor. If you get a reading of 150–900 ohms here, the wiring between terminals Nos. 5 and 6 and the distributor is faulty. If you still do not get a reading between 150–900 ohms, replace the start pick-up. If you received the proper reading when you initially checked terminals Nos. 5 and 6, proceed to the next step.

9. Connect one lead of an ohmmeter to a good engine ground and with the other lead, check the continuity of both start pick-up leads going into the distributor. If there is not continuity, go on to the next step. If you do get a reading, replace the start pick-up.

10. Remove the distributor cap and check the air gap of the start pick-up coil. Adjust, if necessary, and proceed to the next step.

11. Replace the distributor cap, and start the engine; if it still will not start, replace the Spark Control Computer. If the engine still does not work, put the old one back and retrace your steps, paying close attention to any wiring which may be shorted.

ENGINE RUNNING BADLY (RUN PICK-UP TESTS)

1. Start the engine and let it run for a couple of minutes. Disconnect the start pick-up lead. If the engine still runs, leave this test and go on to the "Start Timer Advance Test." If the engine died, proceed to Step 2.

2. Reconnect the start pick-up, turn the ignition switch off, and disconnect the dual connector from the bottom of the spark control computer.

3. Using an ohmmeter, measure the resistance between terminals Nos. 3 and 5 of the dual connector. Resistance should be 150–900 ohms; if it is, proceed to the next step, if not, disconnect the run pick-up leads from the distributor. Measure the resistance going into the distributor. If the resistance is now between 150–900 ohms, there is bad wiring between terminals Nos. 3 and 5 of the double connector plug and the distributor

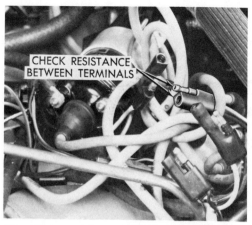

Checking run pick-up at distributor leads

connector terminal. If the resistance is still not within 150–900 ohms, replace the run pick-up, and try to start the engine. If the engine still fails to start, go on to Step 4.

4. Disconnect the run pick-up coil from the distributor. Use an ohmmeter to check for continuity at each of the leads going into the distributor. If there is continuity shown, replace the pick-up coil and repeat Step 1. If you do not get a reading of continuity, proceed to the next step.

5. Remove the distributor cap, check the gap of the run pick-up and adjust it if necessary.

6. Reinstall the distributor cap, check the wiring, and try to start the engine. If it does not start, replace the computer and try again. If it still does not start, repeat the test paying close attention to all wiring connections.

Start Timer Advance Test

1. Hook up an *adjustable* timing light to the engine.

2. Have an assistant start the engine, place his foot firmly on the brake, then open and close the throttle, then place the transmission in Drive.

3. Locate the timing signal immediately after the transmission is put in Drive. The meter on the timing light should show about 5–9° advance. This advance should slowly decrease to the basic timing signal after about one minute. If it did not increase the 5–9° or return after one minute, replace the spark control computer. If it did operate properly, proceed to the next test.

Throttle Advance Test

Before performing this test, the throttle position transducer must be adjusted. The adjustments are as follows:

1. The air temperature sensor inside the spark control computer must be cool (below 135°). If the engine is at operating temperature, either turn it off and let it cool down or remove the top of the air cleaner and inject a spray coolant into the computer over the air temperature sensor for about 15 seconds. If Steps 2–5 take longer than 3–4 minutes, recool the sensor.

2. Start the engine and wait about 90 seconds, then connect a jumper wire between the carburetor switch terminal and ground.

3. Disconnect the electrical connector from the transducer and check the timing, adjusting if necessary. Reconnect the electrical connector to the transducer and recheck the timing.

4. If the timing is more than specified on the tune-up decal, loosen the transducer locknut and turn the transducer clockwise until it comes within limits, then turn it an additional ½ turn clockwise and tighten the locknut.

5. If the timing is at the specified limits, loosen the locknut and turn the transducer counterclockwise until the timing just begins to advance. At that point, turn the transducer ½ turn clockwise and tighten the locknut. After this step you are ready to begin the throttle advance test.

6. Turn the ignition switch "Off" and disconnect the single connector from the bottom of the spark control computer.

7. With an ohmmeter, measure the resistance between terminals Nos. 9 and 10 of the single connector. The measured resistance should be between 50–90 ohms. If it is, go on to the next step; if not, remove the connector from the throttle position transducer and measure the resistance at the transducer terminals. If you now get a reading of 50–90 ohms, check the wiring between the computer terminal and the transducer terminal. If you do not get the 50–90 reading, replace the transducer and proceed to the next step.

8. Reconnect the wiring and turn the switch to the "Run" position without

Checking for resistance at throttle transducer terminals

starting the engine. Hook up a voltmeter, negative lead to an engine ground, and touch the positive lead to one terminal of the transducer while opening and closing the throttle all the way. Do the same thing to the other terminal of the transducer. Both terminals should show a 2 volt change when opening and closing the throttle. If not proceed to the next step.

9. Position the throttle linkage on the fast idle cam and ground the carb switch with a jumper wire. Disconnect the wiring connector from the transducer and connect it to a transducer that you know is good.

10. Move the core of the transducer all the way in, start the engine, wait about 90 seconds, and then move the core out about an inch.

11. Adjust the timing light so that it registers the basic riming signal. The timing light should show the additional amount of advance as given in the "Transducer Advance Specifications"

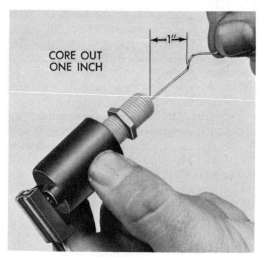

Checking with test transducer

chart in this section. If it is within the specifications, move the core back into the transducer, and the timing should go back to the original position. If the timing did advance and return, go on to the next step. If it did not advance and/or return, replace the spark control computer and try this test over again. If it still fails, replace the transducer.

12. Reset the timing light meter, and have an assistant move the transducer core in and out 5–6 times quickly. The timing should advance 7–12° for about a second and then return to the base figure. If it did not, replace the spark control computer. If you did not get the 2 volt change in reading in Step 8, you should now replace the transducer since you have proved that the spark control computer is not causing it to check out faulty.

13. Remove the test transducer (from Step 9), and reconnect all wiring.

Vacuum Advance Test (Vacuum Transducer)

1. Hook up an adjustable timing light.
2. Turn the ignition switch to the "Run" position, but do not start the engine. Disconnect the idle stop solenoid wire and the wiring connector from the coolant switch. Push the solenoid plunger in all the way, and while holding the throttle linkage open, reconnect the solenoid wire. The solenoid plunger should pop out and when the throttle linkage is released; it should also hold the linkage in place. If it does not, replace the idle stop solenoid.
3. Start the engine and let it warm up; make sure that the transmission is in Neutral and the parking brake is on.
4. Place a small piece of plastic or paper between the carburetor switch and the curb idle adjusting screw; if the screw is not touching the switch, make sure that the fast idle cam is not on or binding; the linkage is not binding, or the throttle stop screw is not overadjusted. Adjust the timing light for the basic timing figure. The meter of the timing light should show 2–5° of advance with a minimum of 16 in. of vacuum at the vacuum transducer (check with a vacuum gauge). If this advance is not present, replace the spark control computer and try the test again. If the advance is present, let the

engine run for about 6–9 minutes, then go on to the next step.

5. After the 6–9 minute waiting period, adjust the timing light so that it registers the basic timing figure. The timing light meter should now register 32–35° of advance. If the advance is not shown, replace the spark control computer and repeat the test; if it is shown, proceed to Step 6.

6. Remove the insulator (paper or plastic) which was installed in Step 4; the timing should return to its base setting. If it does not, make sure that the curb idle adjusting screw is not touching the carburetor switch. If that is alright, turn the engine off and check the wire between terminal No. 11 of the single connector (from the bottom of the spark control computer), and the carburetor switch terminal for a bad connection. If it turns out alright, and the timing still will not return to its base setting, replace the spark control computer.

Coolant Switch Test

1. Connect one lead of the ohmmeter to a good engine ground, the other to the black wire with a tracer in it. Disregard the orange wire if there is one on the switch.
2. If the engine is cold (below 150°), there should be continuity present in the switch. With the thermostat open, and the engine warmed up, there should be no continuity. If either of the conditions in this step are not met, replace the switch.

Lean Burn Timing

This procedure is to set the basic timing signal as shown on the engine tune-up decal in the engine compartment.

1. Connect a jumper wire between the carburetor switch terminal and the ground. Connect a standard timing light to the No. 1 cylinder.
2. Block the wheels and set the parking brake. If the car has an automatic release type parking brake, remove and plug the vacuum line which controls it from the fitting on the rear of the engine.
3. Start and warm the engine up; raise engine speed above 1,500 rpm for a second, then drop the speed and let it idle for a minute or two.
4. With the engine idling at the point

specified on the tune-up decal with the transmission in Drive, adjust the timing to the figure given on the tune-up decal.

Idle Speed and Mixture

1. Follow the first three steps under "Lean Burn Timing," then insert an exhaust gas analyzer into the tailpipe.

2. Place the transmission in Drive, the air conditioning and headlights off. Adjust the idle speed to the specification shown on the tune-up decal by turning the idle solenoid speed screw.

3. Adjust the carbon monoxide level to 0.1% with the mixture screws while trying to keep hydrocarbons to a minimum and the idle speed to specification. Turn the screws alternately over their range to coordinate all three factors.

4. Place the transmission in Neutral; disconnect the wire at the idle stop solenoid and adjust the curb idle speed screw to obtain 650 rpm. Reconnect the wire.

5. Remove the air cleaner cover and lift and support the air cleaner assembly high enough to gain access to the fast idle adjustment screw.

6. Place the fast idle speed screw on the highest step of the cam and adjust the idle speed to the specification shown on the tune-up decal.

7. Drop the idle back down to the curb idle position and turn the ignition switch off. Reconnect any hoses or electrical connections taken off in the procedure.

8. If the procedure has to be performed a second time, make sure that you start from the beginning or the readings will be inaccurate.

Removal and Overhaul

None of the components of the Lean Burn System (except the carburetor), are able to be disassembled and repaired. When one part is known to be defective, it is replaced.

The Spark Control Computer is secured by mounting screws inside the air

Transducer Advance Specifications

Core Moved Out 1 in.	7–12° @ 75° F
	4–7° @ 104° F
Moved 5–6 Times	7–12°
	(one second duration each time)

cleaner. To remove the Throttle position transducer, loosen the locknut and unscrew it from the mounting bracket, then unsnap the core from the carburetor linkage.

PICK-UP GAPS

Start Pick-Up	(set to)	0.008 in.
	(check at)	0.010 in.
Run Pick-Up	(set to)	0.012 in.
	(check at)	0.014 in.

FUEL SYSTEM

Mechanical Fuel Pump
REMOVAL AND INSTALLATION

1. Remove the fuel inlet and outlet lines. Plug the fuel inlet line to prevent emptying the fuel tank.

2. Remove the two fuel pump securing bolts.

3. Remove the fuel pump from the car.

4. Reverse the above steps to install.

Carburetors
REMOVAL AND INSTALLATION

1. Remove the air cleaner assembly.

2. Disconnect the throttle linkage.

3. Note the position of all the vacuum lines and electrical leads which run to the carburetor and disconnect them.

4. Disconnect the choke linkage.

5. Unfasten the carburetor securing nuts and remove the carburetor.

6. Reverse the above steps to install the carburetor. Use a *new* gasket. Tighten the mounting nuts evenly so that the carburetor fits snugly on the manifold.

Make sure that it is not cocked. Adjust the throttle linkage.

THROTTLE LINKAGE ADJUSTMENTS
1968–73 Automatic Transmission
V8—Single Section Throttle Rod

1. Apply a thin film of grease to the friction points of the throttle linkage.

2. Disconnect the choke and make sure the fast idle cam is not holding the throttle open.

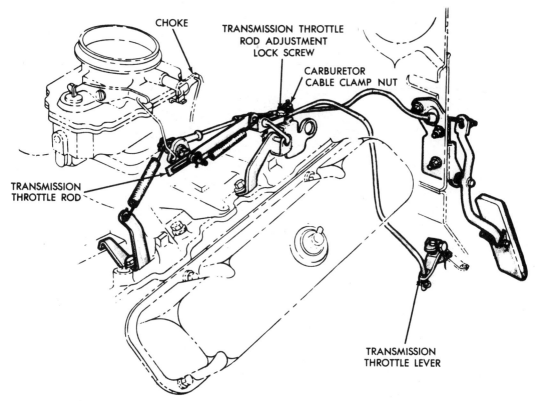

CHOKE

TRANSMISSION THROTTLE
ROD ADJUSTMENT
LOCK SCREW

CARBURETOR
CABLE CLAMP NUT

TRANSMISSION
THROTTLE ROD

TRANSMISSION
THROTTLE LEVER

V8 automatic linkage used with V8 single section throttle rod

3. Insert a ³/₁₆ in. rod into the holes provided in the upper bellcrank and lever. Adjust the length of the intermediate transmission rod by means of the threaded adjustment at the upper end of the rod. The ball socket must line up with the ball end when the rod is held up and the transmission lever is forward against its stop.

4. Assemble the ball socket to the ball end and remove the ³/₁₆ in. rod.

5. Disconnect the return spring. Adjust the length of the carburetor rod by pulling forward on the rod and turning the threaded adjuster link until the rear end of the slot just contacts the carburetor lever pin. Lengthen the rod by two full turns of the link and reinstall the link on the lever pin.

6. Loosen the cable clamp nut. Adjust the position of the cable mounting ferrule in the clamp so that all slack is removed from the cable. To remove the slack, move the ferrule in the clamp away from the carburetor lever. Leave ¼ in. free-play between the front edge of the accelerator shaft lever and the dash bracket.

1968–73 Automatic V8—Three-Section Throttle Rod

1. Lubricate the friction points in the throttle linkage.

2. Disconnect the choke at the carburetor and block it open. Make sure the throttle is off the fast idle cam.

3. Loosen the transmission throttle rod adjustment lockscrew.

4. Hold the transmission lever forward against its stop while adjusting the linkage.

5. Adjust the transmission rod at the carburetor by pushing forward on the retainer and rearward on the rod to remove all slack. Tighten the transmission rod adjustment locking screw. The rear edge of the link slot must be against the carburetor lever pin during this adjustment. Reconnect the choke.

6 Cylinder Automatic—All

1. Lubricate the friction points of the throttle linkage.

2. Disconnect the choke at the carburetor and make sure the throttle is off the fast idle cam.

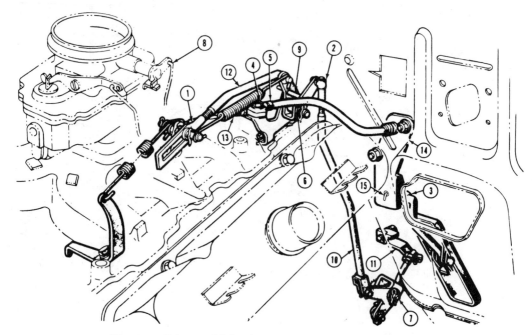

V8 automatic/manual linkage used with three-section throttle rod

1. Adjuster link
2. Ball socket *
3. Accelerator shaft
4. Cable clamping nut
5. Cable housing ferrule

6. Lubrication point
7. Lubrication point *
8. Choke rod
9. 0.188 in. rod *
10. Transmission rod *

11. Transmission lever *
12. Carburetor rod
13. Return spring
14. Lubrication pocket
15. Anti-rattle spring

* Automatic transmission only

3. The transmission lever must remain firmly against its stop while adjusting the throttle linkage.

4. Loosen the slotted link lockbolt to adjust the length of the transmission rod. Pull forward on the slotted adjuster link to maintain pressure against the carburetor lever pin and remove all slack in the linkage.

5. Tighten the transmission rod adjustment lockbolt.

6. To adjust the throttle cable, loosen the cable clamp nut and position the cable up or down to obtain 1/4 in. of slack and then tighten the nut.

7. Reconnect the choke linkage and check the linkage for freedom of operation.

1974–76 V8 Automatic

Adjust the idle speed, prior to performing this adjustment. See Chapter 2 for the correct procedure.

1. Start the engine and allow it to reach normal operating temperature. Be sure that the fast idle screw is not on the fast idle cam.

2. Raise the car and make sure that it is securely supported with jackstands.

3. Loosen the adjustment swivel lock screw on the transmission throttle lever.

4. See that the swivel is free to slide along the flat end of the throttle rod, or the preload spring will be restricted.

5. Move the transmission lever forward, against its internal stop, and tighten its lockscrew to 125 in. lbs while holding it firmly.

NOTE: *Linkage backlash is automatically removed by the preload spring.*

6. Lower the car to the ground and check linkage operation by moving the throttle rod rearward and slowly releasing it. It should return fully forward.

6 Cylinder and V8 Manual Transmission—All

1. Lubricate the friction points of the throttle linkage.

2. Disconnect the choke at the carburetor and make sure the throttle is off the fast idle cam.

3. Loosen the cable clamp nut. Adjust the cable by moving the cable housing so that there is about 1/4 in. of slack in the cable at idle.

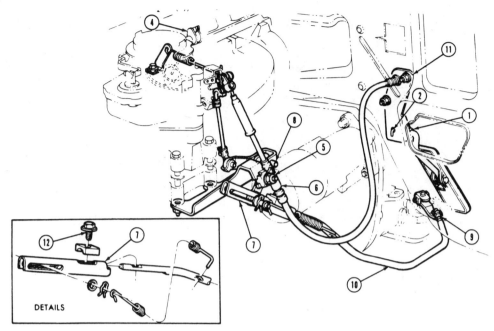

Slant six automatic/manual throttle linkage

1. Lubrication point
2. Anti-rattle spring
3. Lubrication point
4. Choke rod
5. Cable clamping nut
6. Ferrule
7. Adjusting link *
8. Bellcrank pivot pin *
9. Transmission lever *
10. Transmission rod *
11. Lubrication pocket
12. Locking bolt *

* Automatic transmission only

4. Tighten the cable clamp nut.

5. Reconnect the choke and check the linkage for free movement.

FAST IDLE CAM ADJUSTMENT

All—Except Thermo-Quad®

NOTE: *The adjustment for fast idle speed should be made with the carburetor on the car. For the procedure, see below.*

1. With the fast idle-speed adjustment screw touching the second highest step on the fast idle cam, close the choke valve by applying light pressure to it.

2. Insert the specified size of drill or gauge (see the table below) between the top of the choke valve and the wall of the air horn. Withdraw the drill or gauge; a slight interference should be felt. If no interference is felt, adjustment is required.

On some older carburetors (see chart), check to see that the index mark on the cam aligns with the center of the fast idle screw, with both the throttle and the choke valves closed. If it does not, then an adjustment is required.

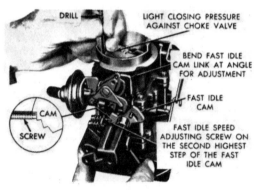

Typical fast idle cam adjustment

3. To adjust the fast idle cam, bend the fast idle connector rod, at its angle, until proper adjustment is obtained.

NOTE: *On Holley 1920 carburetors, the choke unloader is automatically adjusted with the fast idle cam. See below for those carburetors which require a separate unloader adjustment.*

Thermo-Quad®

1. Position the fast idle screw on the second step of the cam so that it rests against the shoulder of the first step.

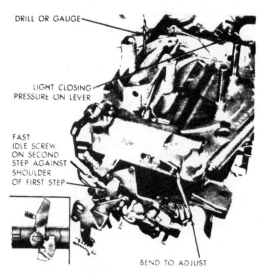

DRILL OR GAUGE

LIGHT CLOSING
PRESSURE ON LEVER

FAST
IDLE SCREW
ON SECOND
STEP AGAINST
SHOULDER
OF FIRST STEP

BEND TO ADJUST

Fast idle adjustment on Thermo-Quad®

2. Bend the fast idle connector rod to obtain a 0.110 in. clearance between the air horn wall and long side of the choke valve nearest the lever.

VACUUM KICK ADJUSTMENT

This adjustment can be performed with the carburetor on or off the engine. Because the choke diaphragm must be energized to measure the vacuum kick adjustment, an auxiliary vacuum source, such as a distributor test machine or vacuum supplied by another vehicle, must be used when the adjustment is performed off the vehicle.

All—Except Thermo-Quad®

1. If the adjustment is to be made *on* the engine (with engine running at curb idle), back off the fast idle screw until the choke can be closed to the kick position. Note the number of screw turns required so that the fast idle can be returned to the original adjustment.
2. If the adjustment is to be made *off* the engine, open the throttle valve and move the choke to its closed position. Release the throttle first and then release the choke. Disconnect the vacuum hose from the carburetor body and connect it to the auxiliary vacuum source. Apply a vacuum of at least 10 in. Hg.
3. Insert the specified size drill or gauge (refer to the appropriate chart) between the choke valve and the wall of the air horn.

4. Apply sufficient closing pressure to the choke lever to provide a minimum valve opening without distorting the diaphragm link (which connects the choke lever to the vacuum diaphragm). Note that the cylindrical stem of the diaphragm will extend as its internal spring is compressed. This spring must be fully compressed for the proper measurement of the vacuum kick adjustment.
5. Remove the drill or gauge. If a slight drag is not felt as the drill or gauge is removed, an adjustment of the diaphragm link is necessary to obtain the proper clearance. Shorten or lengthen the diaphragm link by carefully closing or opening the U-bend in the link until the correct adjustment is obtained.

CAUTION: *When adjusting the link, be careful not to bend or twist the diaphragm.*
6. Refit the vacuum hose to the carburetor body (if it had been removed) and return the fast idle screw to its original location.
7. With *no* vacuum applied to the diaphragm, the choke valve should move freely between its open and closed positions. If it does not move freely, examine the linkage for misalignment or interference which may have been caused by the bending operation. If necessary, repeat the adjustment to provide the proper link operation.

Thermo-Quad®

Before the vacuum kick adjustment can be performed, the choke control lever and the choke diaphragm connector rod must be correctly set. These settings can be made on or off the vehicle.
1. If the setting is to be made *on* the engine, remove the choke assembly, stainless steel cup, and gasket. If the setting is to be *off* the vehicle, place the carburetor on a clean, flat surface, such as a table top or workbench, so that the carburetor flange is flush against the work surface.
2. Close the choke valve by pushing on the choke lever with the throttle partly open.
3. Measure the vertical distance from the top of the rod hole in the choke control lever down to the clean choke pad surface (on engine), or down to the work surface (off engine). This measurement

should be 5.641 in. on the engine, or 3.422 off the engine.

4. If an adjustment is necessary, bend the link which connects the two choke shafts until the correct measurement is obtained.

5. Refit the choke assembly (if it had been removed).

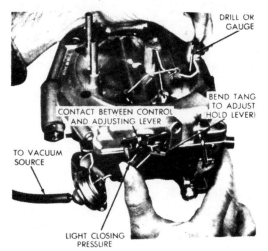

Vacuum kick adjustment on Thermo-Quad®

After the choke control lever has been properly set, continue this procedure to adjust the choke diaphragm connector rod and the vacuum kick. The choke diaphragm must be energized during both of these adjustments.

1. If the adjustment is to be made *on* the engine (with the engine running at curb idle), measure the clearance between the air valve and its stop (see illustration). The clearance should be 0.040 in. with the air valve closed. If necessary, adjust the connector rod as illustrated until the correct clearance is obtained. Then back off the fast idle screw until the choke can be closed to the kick position. Note the number of screw turns required so that the fast idle can be returned to the original adjustment.

2. If the adjustment is to be made *off* the engine, disconnect the vacuum hose from the carburetor body and connect it to the auxiliary vacuum source. Apply a vacuum of at least 15 in. Hg and measure the clearance between the air valve and its stop (see illustration). The clearance should be 0.040 in. with the air valve closed. If necessary, adjust the connector rod as illustrated until the cor-

rect clearance is obtained. Disconnect the vacuum line from the auxiliary source, open the throttle valves, and move the choke valve to its closed position with the control lever. Release the throttle before releasing the choke to trap the fast idle cam in the closed choke position. Reconnect the vacuum line to the auxiliary vacuum source and again apply a vacuum of at least 15 in. Hg.

3. Insert the specified size drill or gauge (refer to the appropriate chart) between the long side (lower edge) of the choke valve and the air horn wall.

4. Apply sufficient closing pressure to the choke control lever to provide a minimum choke valve opening without distorting the choke linkage. Note that only this carburetor extends a spring connecting the control lever to an adjustment lever. This spring must be fully extended for the proper measurement of the vacuum kick adjustment.

5. Remove the drill or gauge. If a slight drag is not felt as the drill or gauge, is removed, an adjustment of the adjusting lever is necessary to obtain the proper clearance. While applying a counterforce to the adjusting lever, bend the adjusting lever tang to change the contact with the end of the diaphragm rod.

CAUTION: *Do not adjust the diaphragm rod. Do not bend the link which connects the two choke shafts because the choke control lever adjustment will be changed.*

6. Refit the vacuum hose to the carburetor body (if it had been removed) and return the fast idle screw to its original position.

7. With *no* vacuum applied to the diaphragm, the choke valve should move freely between its open and closed positions. If it does not move freely, examine the linkage for misalignment or interference which may have been caused by the bending operation. If necessary, repeat the adjustment to provide the proper linkage operation.

UNLOADER ADJUSTMENT
All—Except Holley 1920

The choke unloader partially opens the choke valve at full throttle to prevent choke enrichment during engine crank-

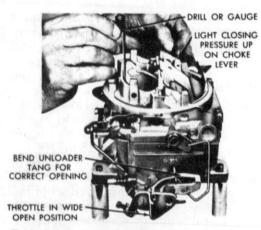

DRILL OR GAUGE

LIGHT CLOSING PRESSURE UP ON CHOKE LEVER

BEND UNLOADER TANG FOR CORRECT OPENING

THROTTLE IN WIDE OPEN POSITION

Typical choke unloader adjustment

ing. Engines that have been stalled or flooded by too much choke enrichment can be started by using the unloader (fully depressing the accelerator pedal).

Unloader adjustment is performed in the following manner:

1. Hold the throttle valve(s) wide open. Insert the proper size drill (see chart below) between the upper edge of the choke valve and the air horn inner wall.

2. Lightly press against the valve and withdraw the drill. A slight drag should be felt. If none is felt, or if there is too much drag, an adjustment is required.

3. To adjust the unloader on all but the Carter AVS and Thermo-Quad carburetors, bend the tang on the throttle lever until the proper opening has been reached.

4. On the Carter AVS and Thermo-Quad, bend the unloader tang on the *fast idle cam* until the proper opening has been reached.

Holley 1920

When the fast idle cam is adjusted, the choke unloader is automatically adjusted and requires no further attention.

FAST IDLE SPEED ADJUSTMENT

All Carburetors

1. Warm up the engine by driving for at least five miles. Connect a tachometer. Make the curb idle and mixture adjustments first (see "Tune-Up"); then proceed with the following steps:

2. Remove the air cleaner and plug the OSAC valve and heated air cleaner

motor vacuum lines (if so equipped). Turn the engine off and place the transmission in Neutral or Park. Open the throttle slightly.

3. Close the choke valve so that the fast idle screw can be placed on the second highest step of the fast idle cam.

4. Start the engine and allow its speed to stabilize. Turn the fast idle screw in or out, as required, to obtain the specified fast idle speed (see chart below).

5. It is unnecessary to stop the engine between adjustments. Always reposition the fast idle speed screw on the fast idle cam after each adjustment, however, to maintain proper throttle closing torque.

NOTE: *Always check the timing for correct adjustment before adjusting the curb idle and fast idle speeds.*

FLOAT AND FUEL LEVEL ADJUSTMENTS

Carter BBD and Rochester 2GV

1. Remove the carburetor from the engine.

2. Remove the clips and disengage the accelerator pump operating rod.

3. Remove the clips and disengage the fast idle connector rod from the fast idle cam and choke lever.

4. Remove the vacuum hose from between the carburetor main body and the choke vacuum diaphragm.

5. Remove the clip from the choke operating link and disengage the link from the diaphragm plunger (stem) and choke lever.

6. Remove the choke vacuum diaphragm and bracket assembly.

7. Remove the air horn attachment screws and lift the air horn straight up, and away from the main body.

8. Invert the main body (catch the

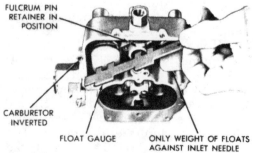

FULCRUM PIN RETAINER IN POSITION

CARBURETOR INVERTED

FLOAT GAUGE   ONLY WEIGHT OF FLOATS AGAINST INLET NEEDLE

Float level adjustment—Carter BBD (Rochester 2GV similar)

pump intake check ball) so that the weight of only the floats is forcing the needle against its seat.

9. Using a gauge of suitable depth, measure the distance from the surface of the fuel bowl to the crown of each float. This measurement should be as specified in the appropriate carburetor specifications chart.

10. If an adjustment is necessary, hold the float(s) on the bottom of the bowl and bend the float lip toward, or away from, the needle. Check the setting and then repeat the lip bending operation as required.

CAUTION: *When bending the float lip, do not allow it to push against the needle, as the synthetic rubber tip can be compressed sufficiently to cause a false setting that will affect the proper level of fuel in the bowl.*

11. After the correct setting has been obtained, reverse steps 1–8 to reassemble the carburetor and install the carburetor on the engine.

Holley 1920

1. Remove the carburetor from the engine.

2. Remove the economizer cover retaining screws, then lift the cover, diaphragm, and stem out of the carburetor.

3. Remove the fuel bowl securing screws and remove the fuel bowl and baffle.

4. Invert the carburetor and slide the float gauge (included in carburetor overhaul kit) into position.

5. The float should just contact the "TOUCH" leg of the gauge. Reverse the gauge and the float should just clear the "NO TOUCH" leg of the gauge.

6. If an adjustment is necessary, bend the float tab using needle-nose pliers.

CAUTION: *Do not allow the float tab to contact the float needle head during the bending operation as the synthetic rubber tip of the needle can be compressed sufficiently to cause a false setting. Do not touch the contact area of the float tab with the pliers.*

7. When the float adjustment is correct, refit the economizer assembly.

8. Be sure that the baffle is in position in the fuel bowl, place the gasket on the bowl, and refit this assembly to the carburetor.

9. Install the bowl securing screws and washers, and tighten the screws gently so that only the washers are compressed and the fuel bowl is not distorted.

Carter AVS

1. Remove the carburetor from the engine.

2. Remove the clip which secures the fast idle connector rod to the choke lever. Disengage the rod from the lever and swing the rod in an arc until it can be disengaged from the fast idle cam.

3. Remove the pin and clip which attach the throttle connector rod to the accelerator pump arm and the primary throttle shaft lever. Disengage the rod from the arm and lever, and remove it from the carburetor.

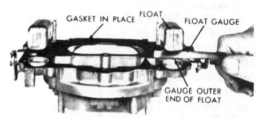

Float level adjustment—Carter AVS

4. Withdraw the securing screws from the step-up piston and rod cover plates.

NOTE: *Hold the cover down while removing the screws to keep the piston and rods from flying out.*

5. Lift off the plates and slide the step-up pistons and rods out of the air horn. Remove the step-up piston springs.

6. Detach the vacuum hose from between the carburetor throttle body and the vacuum diaphragm.

7. Remove the clip from the choke operating link and disengage the link from the diaphragm plunger (stem) and the choke lever.

8. Remove the vacuum diaphragm and bracket assembly.

9. Unfasten the air horn retaining screws and lift the air horn straight up, and away from the main body.

CAUTION: *When removing the air horn, be careful not to bend or damage the floats.*

10. Invert the air horn. With the air horn gasket in place and the float needle seated, slide the proper size feeler gauge

(refer to appropriate carburetor chart) between the top of the float at its outer end and the air horn gasket. The float should just touch the gauge.

11. Check the other float in the same manner. If an adjustment is necessary, bend the float arm until the correct clearance is obtained.

12. After the proper clearance has been obtained, check the float alignment by sighting down the side of each float shell to determine if the side of the float is parallel to the outer cage of the air horn.

13. If the sides of the float are not in alignment with the edge of the casting, bend the float lever by applying thumb pressure to the end of the float shell.

NOTE: *Apply only enough pressure to bend the float lever in order to avoid damaging the float.*

14. After aligning the floats, remove as much clearance as possible from between the arms of the float lever and the lugs of the air horn. To do this, bend the float lever. The arms of the float lever should be as parallel as possible to the inner surfaces of the air horn lugs.

15. When all adjustments are complete, reverse Steps 1–9, to reassemble the carburetor, and install the carburetor on the engine.

Carter Thermo-Quad®

1. Remove the carburetor from the engine.

2. Remove the retainers which secure the throttle connector rod to the accelerator pump arm and throttle lever. Remove the rod from the carburetor.

3. Unfasten the accelerator pump arm screw and disengage the pump rod S-link (leave the S-link connected to the pump rod). Then, remove the lever.

4. Remove the choke countershaft fast-idle lever attachment screw while holding the lever. Disengage the lever from the countershaft and then swing the fast idle connector rod in an arc until it can be disengaged from the fast idle operating lever.

5. Remove the retainers and washer which secure the choke diaphragm connector rod to the choke vacuum diaphragm and air valve lever. Remove the lever.

6. Remove the retainer which at-

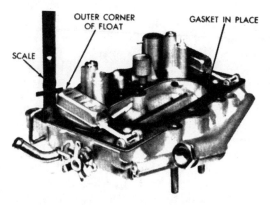

Float level adjustment—Carter Thermo-Quad®

taches the choke connector rod to the choke countershaft. Disengage and swing the rod in an arc to remove the choke shaft lever assembly.

7. Withdraw the step-up piston coverplate securing screw and cover plate. Remove the step-up piston and link assembly with the step-up rods. Remove the step-up piston spring.

8. Remove the pump jet housing screw, housing and gasket. Invert the carburetor and remove the discharge check needle.

9. Withdraw the bowl cover retaining screws and remove the bowl cover.

10. Invert the bowl cover. With the bowl cover gasket in place and the float needle seated, use a depth gauge to measure the distance from the bowl cover gasket to the bottom side of the float. This dimension should be 1.00 in.

11. If an adjustment is necessary, bend the float lever until the correct distance is obtained.

CAUTION: *Never allow the lip of the float to be pressed against the needle when adjusting the float.*

12. When the float adjustment is complete, reverse Steps 1–9, to reassemble the carburetor, and install the carburetor on the engine.

Holley 2210 and 2245

1. Remove the nut and washer attaching the accelerator pump rocker arm to the acclerator pump shaft. Remove the arm from the shaft.

2. Remove the accelerator pump rod from the arm.

3. Remove the choke diaphragm and hose.

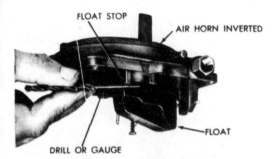

Float level adjustment—Holley 2210 and 2245

4. Remove the choke lever from the choke valve shaft.

5. Remove the eight air horn attaching screws and lift the air horn straight up and away from the carburetor body. Be careful not to damage the float and main well tubes attached to the bottom of the air horn.

6. Invert the air horn and measure the clearance between the top of the float and the float stop.

7. Bend the tang on the float arm until the float is properly adjusted.

8. Reverse Steps 1–5 to install.

Holley 2300

1. Start the engine and remove the sight plug from the fuel bowl.

2. Loosen the adjusting screw locknut and turn the fuel level adjusting screw until fuel just starts to dribble out of the sight hole. Use a rag to catch any fuel that comes out of the sight hole.

3. Reinstall the sight hole plug and tighten the locknut.

Holley 4160

1. Remove the fuel bowl attaching screws and slide the bowl off the fuel transfer tube.

2. Invert the fuel bowl. Measure the clearance between the toe of the float and the surface of the fuel bowl. The measurement for the primary fuel bowl is 0.110 in. and the secondary fuel bowl is 0.204 in. Bend the float tang to obtain the correct clearance.

3. Reinstall the fuel bowl.

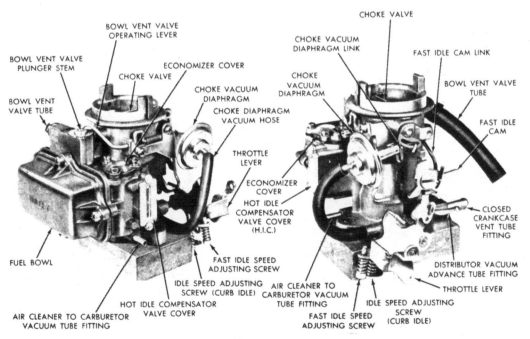

Holley 1929

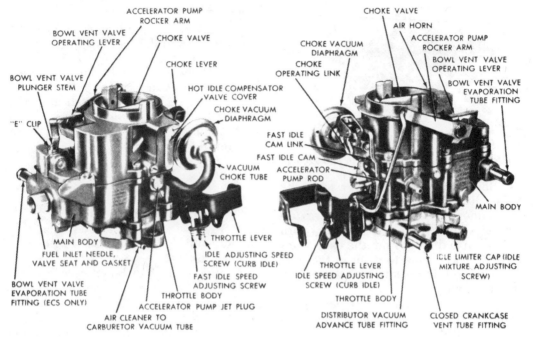

Carter BBS

OVERHAUL

All Types

Efficient carburetion depends greatly on careful cleaning and inspection during overhaul since dirt, gum, water, or varnish in or on the carburetor parts are often responsible for poor performance.

Overhaul your carburetor in a clean, dust-free area. Carefully disassemble the carburetor, referring often to the ex-ploded views. Keep all similar and look-alike parts segregated during disassembly and cleaning to avoid accidental interchange during assembly. Make a note of all jet sizes.

When the carburetor is disassembled, wash all parts (except diaphragms, electric choke units, pump plunger, and any other plastic, leather, fiber, or rubber parts) in clean carburetor solvent. Do not leave parts in the solvent any longer

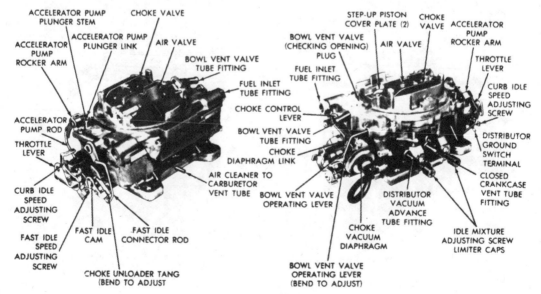

Carter AVS

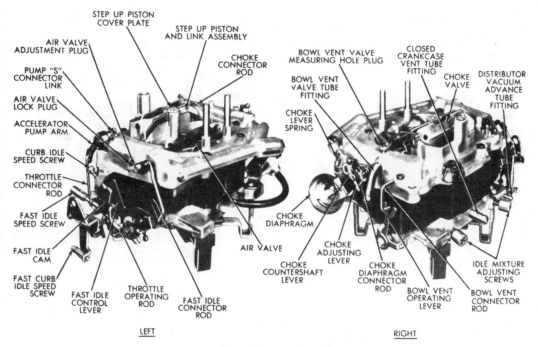

Carter Thermo-Quad®

than is necessary to sufficiently loosen the deposits. Excessive cleaning may remove the special finish from the float bowl and choke valve bodies, leaving these parts unfit for service. Rinse all parts in clean solvent and blow them dry with compressed air or allow them to air dry. Wipe clean all cork, plastic, leather, and fiber parts with a clean, lint-free cloth.

Blow out all passages and jets with compressed air and be sure that there are no restrictions or blockages. Never use wire or similar tools to clean jets,

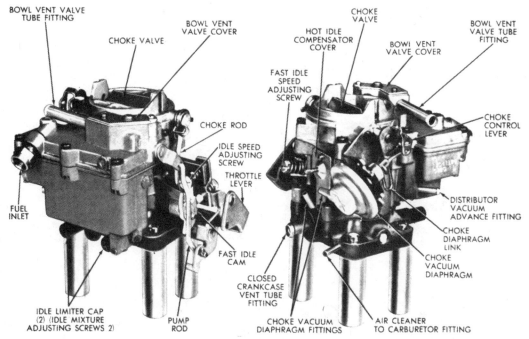

Rochester 2GV

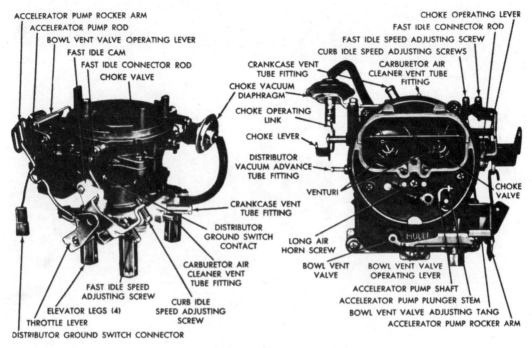

ACCELERATOR PUMP ROCKER ARM
ACCELERATOR PUMP ROD
BOWL VENT VALVE OPERATING LEVER
FAST IDLE CAM
FAST IDLE CONNECTOR ROD
CHOKE VALVE
CRANKCASE VENT TUBE FITTING
CHOKE VACUUM DIAPHRAGM
CHOKE OPERATING LINK
CHOKE LEVER
DISTRIBUTOR VACUUM ADVANCE TUBE FITTING
VENTURI
CRANKCASE VENT TUBE FITTING
DISTRIBUTOR GROUND SWITCH CONTACT
LONG AIR HORN SCREW
CARBURETOR AIR CLEANER VENT TUBE FITTING
FAST IDLE SPEED ADJUSTING SCREW
ELEVATOR LEGS (4)
THROTTLE LEVER
CURB IDLE SPEED ADJUSTING SCREW
DISTRIBUTOR GROUND SWITCH CONNECTOR

CHOKE OPERATING LEVER
FAST IDLE CONNECTOR ROD
FAST IDLE SPEED ADJUSTING SCREW
CURB IDLE SPEED ADJUSTING SCREWS
CARBURETOR AIR CLEANER VENT TUBE FITTING
CHOKE VALVE
BOWL VENT VALVE
BOWL VENT VALVE OPERATING LEVER
ACCELERATOR PUMP SHAFT
ACCELERATOR PUMP PLUNGER STEM
BOWL VENT VALVE ADJUSTING TANG
ACCELERATOR PUMP ROCKER ARM

Holley 2210 (2245 similar)

fuel passages, or air bleeds. Clean all jets and valves separately to avoid accidental interchange.

Check all parts for wear or damage. If wear or damage is found, replace the defective parts. Especially check the following:

1. Check the float needle and seat for wear. If wear is found, replace the complete assembly.

2. Check the float hinge pin for wear and the float(s) for dents or distortion. Replace the float if fuel has leaked into it.

3. Check the throttle and choke shaft

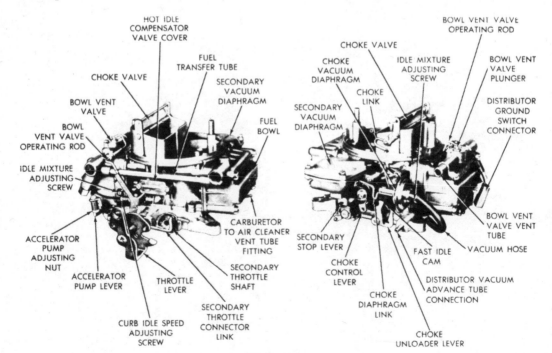

HOT IDLE COMPENSATOR VALVE COVER
FUEL TRANSFER TUBE
CHOKE VALVE
SECONDARY VACUUM DIAPHRAGM
BOWL VENT VALVE
BOWL VENT VALVE OPERATING ROD
FUEL BOWL
IDLE MIXTURE ADJUSTING SCREW
ACCELERATOR PUMP ADJUSTING NUT
ACCELERATOR PUMP LEVER
THROTTLE LEVER
CARBURETOR TO AIR CLEANER VENT TUBE FITTING
SECONDARY THROTTLE SHAFT
CURB IDLE SPEED ADJUSTING SCREW
SECONDARY THROTTLE CONNECTOR LINK

BOWL VENT VALVE OPERATING ROD
CHOKE VALVE
CHOKE VACUUM DIAPHRAGM
IDLE MIXTURE ADJUSTING SCREW
BOWL VENT VALVE PLUNGER
CHOKE LINK
SECONDARY VACUUM DIAPHRAGM
DISTRIBUTOR GROUND SWITCH CONNECTOR
SECONDARY STOP LEVER
CHOKE CONTROL LEVER
FAST IDLE CAM
CHOKE DIAPHRAGM LINK
BOWL VENT VALVE VENT TUBE
VACUUM HOSE
DISTRIBUTOR VACUUM ADVANCE TUBE CONNECTION
CHOKE UNLOADER LEVER

Holley 4160

CHILTON'S
FUEL ECONOMY
& TUNE-UP TIPS

Tune-Up • Spark Plug Diagnosis • Emission Controls

Fuel System • Cooling System • Tires and Wheels

General Maintenance

55 WAYS TO IMPROVE FUEL ECONOMY

CHILTON'S FUEL ECONOMY & TUNE-UP TIPS

Fuel economy is important to everyone, no matter what kind of vehicle you drive. The maintenance-minded motorist can save both money and fuel using these tips and the periodic maintenance and tune-up procedures in this Repair and Tune-Up Guide.

There are more than 130,000,000 cars and trucks registered for private use in the United States. Each travels an average of 10-12,000 miles per year, and, in total they consume close to 70 billion gallons of fuel each year. This represents nearly ⅔ of the oil imported by the United States each year. The Federal government's goal is to reduce consumption 10% by 1985. A variety of methods are either already in use or under serious consideration, and they all affect your driving and the cars you will drive. In addition to "down-sizing", the auto industry is using or investigating the use of electronic fuel delivery, electronic engine controls and alternative engines for use in smaller and lighter vehicles, among other alternatives to meet the federally mandated Corporate Average Fuel Economy (CAFE) of 27.5 mpg by 1985. The government, for its part, is considering rationing, mandatory driving curtailments and tax increases on motor vehicle fuel in an effort to reduce consumption. The government's goal of a 10% reduction could be realized — and further government regulation avoided — if every private vehicle could use just 1 less gallon of fuel per week.

How Much Can You Save?

Tests have proven that almost anyone can make at least a 10% reduction in fuel consumption through regular maintenance and tune-ups. When a major manufacturer of spark plugs sur-

TUNE-UP

1. Check the cylinder compression to be sure the engine will really benefit from a tune-up and that it is capable of producing good fuel economy. A tune-up will be wasted on an engine in poor mechanical condition.

2. Replace spark plugs regularly. New spark plugs alone can increase fuel economy 3%.

3. Be sure the spark plugs are the correct type (heat range) for your vehicle. See the Tune-Up Specifications.

Heat range refers to the spark plug's ability to conduct heat away from the firing end. It must conduct the heat away in an even pattern to avoid becoming a source of pre-ignition, yet it must also operate hot enough to burn off conductive deposits that could cause misfiring.

The heat range is usually indicated by a number on the spark plug, part of the manufacturer's designation for each individual spark plug. The numbers in bold-face indicate the heat range in each manufacturer's identification system.

Periodically, check the spark plugs to be sure they are firing efficiently. They are excellent indicators of the internal condition of your engine.

Manufacturer	Typical Designation
AC	R **45** TS
Bosch (old)	WA **145** T30
Bosch (new)	HR **8** Y
Champion	RBL **15** Y
Fram/Autolite	**415**
Mopar	P-**62** PR
Motorcraft	BRF-**42**
NGK	BP **5** ES-15
Nippondenso	W **16** EP
Prestolite	14GR **5** 2A

On AC, Bosch (new), Champion, Fram/Autolite, Mopar, Motorcraft and Prestolite, a higher number indicates a hotter plug. On Bosch (old), NGK and Nippondenso, a higher number indicates a colder plug.

4. Make sure the spark plugs are properly gapped. See the Tune-Up Specifications in this book.

5. Be sure the spark plugs are firing efficiently. The illustrations on the next 2 pages show you how to "read" the firing end of the spark plug.

6. Check the ignition timing and set it to specifications. Tests show that almost all cars

veyed over 6,000 cars nationwide, they found that a tune-up, on cars that needed one, increased fuel economy over 11%. Replacing worn plugs alone, accounted for a 3% increase. The same test also revealed that 8 out of every 10 vehicles will have some maintenance deficiency that will directly affect fuel economy, emissions or performance. Most of this mileage-robbing neglect could be prevented with regular maintenance.

Modern engines require that all of the functioning systems operate properly for maximum efficiency. A malfunction anywhere wastes fuel. You can keep your vehicle running as efficiently and economically as possible, by being aware of your vehicles operating and performance characteristics. If your vehicle suddenly develops performance or fuel economy problems it could be due to one or more of the following:

PROBLEM	POSSIBLE CAUSE
Engine Idles Rough	Ignition timing, idle mixture, vacuum leak or something amiss in the emission control system.
Hesitates on Acceleration	Dirty carburetor or fuel filter, improper accelerator pump setting, ignition timing or fouled spark plugs.
Starts Hard or Fails to Start	Worn spark plugs, improperly set automatic choke, ice (or water) in fuel system.
Stalls Frequently	Automatic choke improperly adjusted and possible dirty air filter or fuel filter.
Performs Sluggishly	Worn spark plugs, dirty fuel or air filter, ignition timing or automatic choke out of adjustment.

Check spark plug wires on conventional point type ignition for cracks by bending them in a loop around your finger.

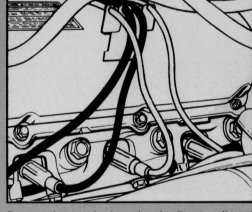

Be sure that spark plug wires leading to adjacent cylinders do not run too close together. (Photo courtesy Champion Spark Plug Co.)

have incorrect ignition timing by more than 2°.

7. If your vehicle does not have electronic ignition, check the points, rotor and cap as specified.

8. Check the spark plug wires (used with conventional point-type ignitions) for cracks and burned or broken insulation by bending them in a loop around your finger. Cracked wires decrease fuel efficiency by failing to deliver full voltage to the spark plugs. One misfiring spark plug can cost you as much as 2 mpg.

9. Check the routing of the plug wires. Misfiring can be the result of spark plug leads to adjacent cylinders running parallel to each other and too close together. One wire tends to pick up voltage from the other causing it to fire "out of time".

10. Check all electrical and ignition circuits for voltage drop and resistance.

11. Check the distributor mechanical and/or vacuum advance mechanisms for proper functioning. The vacuum advance can be checked by twisting the distributor plate in the opposite direction of rotation. It should spring back when released.

12. Check and adjust the valve clearance on engines with mechanical lifters. The clearance should be slightly loose rather than too tight.

SPARK PLUG DIAGNOSIS

Normal

APPEARANCE: This plug is typical of one operating normally. The insulator nose varies from a light tan to grayish color with slight electrode wear. The presence of slight deposits is normal on used plugs and will have no adverse effect on engine performance. The spark plug heat range is correct for the engine and the engine is running normally.

CAUSE: Properly running engine.

RECOMMENDATION: Before reinstalling this plug, the electrodes should be cleaned and filed square. Set the gap to specifications. If the plug has been in service for more than 10-12,000 miles, the entire set should probably be replaced with a fresh set of the same heat range.

Oil Deposits

APPEARANCE: The firing end of the plug is covered with a wet, oily coating.

CAUSE: The problem is poor oil control. On high mileage engines, oil is leaking past the rings or valve guides into the combustion chamber. A common cause is also a plugged PCV valve, and a ruptured fuel pump diaphragm can also cause this condition. Oil fouled plugs such as these are often found in new or recently overhauled engines, before normal oil control is achieved, and can be cleaned and reinstalled.

RECOMMENDATION: A hotter spark plug may temporarily relieve the problem, but the engine is probably in need of work.

Incorrect Heat Range

APPEARANCE: The effects of high temperature on a spark plug are indicated by clean white, often blistered insulator. This can also be accompanied by excessive wear of the electrode, and the absence of deposits.

CAUSE: Check for the correct spark plug heat range. A plug which is too hot for the engine can result in overheating. A car operated mostly at high speeds can require a colder plug. Also check ignition timing, cooling system level, fuel mixture and leaking intake manifold.

RECOMMENDATION: If all ignition and engine adjustments are known to be correct, and no other malfunction exists, install spark plugs one heat range colder.

Photos Courtesy Champion Spark Plug Co.

Carbon Deposits

APPEARANCE: Carbon fouling is easily identified by the presence of dry, soft, black, sooty deposits.

CAUSE: Changing the heat range can often lead to carbon fouling, as can prolonged slow, stop-and-start driving. If the heat range is correct, carbon fouling can be attributed to a rich fuel mixture, sticking choke, clogged air cleaner, worn breaker points, retarded timing or low compression. If only one or two plugs are carbon fouled, check for corroded or cracked wires on the affected plugs. Also look for cracks in the distributor cap between the towers of affected cylinders.

RECOMMENDATION: After the problem is corrected, these plugs can be cleaned and reinstalled if not worn severely.

MMT Fouled

APPEARANCE: Spark plugs fouled by MMT (Methycyclopentadienyl Maganese Tricarbonyl) have reddish, rusty appearance on the insulator and side electrode.

CAUSE: MMT is an anti-knock additive in gasoline used to replace lead. During the combustion process, the MMT leaves a reddish deposit on the insulator and side electrode.

RECOMMENDATION: No engine malfunction is indicated and the deposits will not affect plug performance any more than lead deposits (see Ash Deposits). MMT fouled plugs can be cleaned, regapped and reinstalled.

High Speed Glazing

APPEARANCE: Glazing appears as shiny coating on the plug, either yellow or tan in color.

CAUSE: During hard, fast acceleration, plug temperatures rise suddenly. Deposits from normal combustion have no chance to fluff-off; instead, they melt on the insulator forming an electrically conductive coating which causes misfiring.

RECOMMENDATION: Glazed plugs are not easily cleaned. They should be replaced with a fresh set of plugs of the correct heat range. If the condition recurs, using plugs with a heat range one step colder may cure the problem.

Ash (Lead) Deposits

APPEARANCE: Ash deposits are characterized by light brown or white colored deposits crusted on the side or center electrodes. In some cases it may give the plug a rusty appearance.

CAUSE: Ash deposits are normally derived from oil or fuel additives burned during normal combustion. Normally they are harmless, though excessive amounts can cause misfiring. If deposits are excessive in short mileage, the valve guides may be worn.

RECOMMENDATION: Ash-fouled plugs can be cleaned, gapped and reinstalled.

Detonation

APPEARANCE: Detonation is usually characterized by a broken plug insulator.

CAUSE: A portion of the fuel charge will begin to burn spontaneously, from the increased heat following ignition. The explosion that results applies extreme pressure to engine components, frequently damaging spark plugs and pistons.

Detonation can result by over-advanced ignition timing, inferior gasoline (low octane) lean air/fuel mixture, poor carburetion, engine lugging or an increase in compression ratio due to combustion chamber deposits or engine modification.

RECOMMENDATION: Replace the plugs after correcting the problem.

EMISSION CONTROLS

13. Be aware of the general condition of the emission control system. It contributes to reduced pollution and should be serviced regularly to maintain efficient engine operation.

14. Check all vacuum lines for dried, cracked or brittle conditions. Something as simple as a leaking vacuum hose can cause poor performance and loss of economy.

15. Avoid tampering with the emission control system. Attempting to improve fuel econ-

FUEL SYSTEM

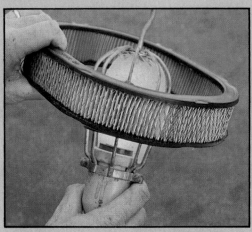

Check the air filter with a light behind it. If you can see light through the filter it can be reused.

Extremely clogged filters should be discarded and replaced with a new one.

18. Replace the air filter regularly. A dirty air filter richens the air/fuel mixture and can increase fuel consumption as much as 10%. Tests show that ⅓ of all vehicles have air filters in need of replacement.

19. Replace the fuel filter at least as often as recommended.

20. Set the idle speed and carburetor mixture to specifications.

21. Check the automatic choke. A sticking or malfunctioning choke wastes gas.

22. During the summer months, adjust the automatic choke for a leaner mixture which will produce faster engine warm-ups.

COOLING SYSTEM

29. Be sure all accessory drive belts are in good condition. Check for cracks or wear.

30. Adjust all accessory drive belts to proper tension.

31. Check all hoses for swollen areas, worn spots, or loose clamps.

32. Check coolant level in the radiator or ex-pansion tank.

33. Be sure the thermostat is operating properly. A stuck thermostat delays engine warm-up and a cold engine uses nearly twice as much fuel as a warm engine.

34. Drain and replace the engine coolant at least as often as recommended. Rust and scale

TIRES & WHEELS

38. Check the tire pressure often with a pencil type gauge. Tests by a major tire manufacturer show that 90% of all vehicles have at least 1 tire improperly inflated. Better mileage can be achieved by over-inflating tires, but never exceed the maximum inflation pressure on the side of the tire.

39. If possible, install radial tires. Radial tires deliver as much as ½ mpg more than bias belted tires.

40. Avoid installing super-wide tires. They only create extra rolling resistance and decrease fuel mileage. Stick to the manufacturer's recommendations.

41. Have the wheels properly balanced.

omy by tampering with emission controls is more likely to worsen fuel economy than improve it. Emission control changes on modern engines are not readily reversible.

16. Clean (or replace) the EGR valve and lines as recommended.

17. Be sure that all vacuum lines and hoses are reconnected properly after working under the hood. An unconnected or misrouted vacuum line can wreak havoc with engine performance.

23. Check for fuel leaks at the carburetor, fuel pump, fuel lines and fuel tank. Be sure all lines and connections are tight.

24. Periodically check the tightness of the carburetor and intake manifold attaching nuts and bolts. These are a common place for vacuum leaks to occur.

25. Clean the carburetor periodically and lubricate the linkage.

26. The condition of the tailpipe can be an excellent indicator of proper engine combustion. After a long drive at highway speeds, the inside of the tailpipe should be a light grey in color. Black or soot on the insides indicates an overly rich mixture.

27. Check the fuel pump pressure. The fuel pump may be supplying more fuel than the engine needs.

28. Use the proper grade of gasoline for your engine. Don't try to compensate for knocking or "pinging" by advancing the ignition timing. This practice will only increase plug temperature and the chances of detonation or pre-ignition with relatively little performance gain.

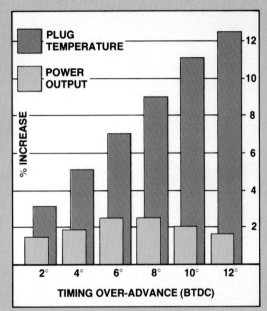

Increasing ignition timing past the specified setting results in a drastic increase in spark plug temperature with increased chance of detonation or preignition. Performance increase is considerably less. (Photo courtesy Champion Spark Plug Co.)

that form in the engine should be flushed out to allow the engine to operate at peak efficiency.

35. Clean the radiator of debris that can decrease cooling efficiency.

36. Install a flex-type or electric cooling fan, if you don't have a clutch type fan. Flex fans use curved plastic blades to push more air at low speeds when more cooling is needed; at high speeds the blades flatten out for less resistance. Electric fans only run when the engine temperature reaches a predetermined level.

37. Check the radiator cap for a worn or cracked gasket. If the cap does not seal properly, the cooling system will not function properly.

42. Be sure the front end is correctly aligned. A misaligned front end actually has wheels going in different directions. The increased drag can reduce fuel economy by .3 mpg.

43. Correctly adjust the wheel bearings. Wheel bearings that are adjusted too tight increase rolling resistance.

Check tire pressures regularly with a reliable pocket type gauge. Be sure to check the pressure on a cold tire.

GENERAL MAINTENANCE

Check the fluid levels (particularly engine oil) on a regular basis. Be sure to check the oil for grit, water or other contamination.

A vacuum gauge is another excellent indicator of internal engine condition and can also be installed in the dash as a mileage indicator.

44. Periodically check the fluid levels in the engine, power steering pump, master cylinder, automatic transmission and drive axle.

45. Change the oil at the recommended interval and change the filter at every oil change. Dirty oil is thick and causes extra friction between moving parts, cutting efficiency and increasing wear. A worn engine requires more frequent tune-ups and gets progressively worse fuel economy. In general, use the lightest viscosity oil for the driving conditions you will encounter.

46. Use the recommended viscosity fluids in the transmission and axle.

47. Be sure the battery is fully charged for fast starts. A slow starting engine wastes fuel.

48. Be sure battery terminals are clean and tight.

49. Check the battery electrolyte level and add distilled water if necessary.

50. Check the exhaust system for crushed pipes, blockages and leaks.

51. Adjust the brakes. Dragging brakes or brakes that are not releasing create increased drag on the engine.

52. Install a vacuum gauge or miles-per-gallon gauge. These gauges visually indicate engine vacuum in the intake manifold. High vacuum = good mileage and low vacuum = poorer mileage. The gauge can also be an excellent indicator of internal engine conditions.

53. Be sure the clutch is properly adjusted. A slipping clutch wastes fuel.

54. Check and periodically lubricate the heat control valve in the exhaust manifold. A sticking or inoperative valve prevents engine warm-up and wastes gas.

55. Keep accurate records to check fuel economy over a period of time. A sudden drop in fuel economy may signal a need for tune-up or other maintenance.

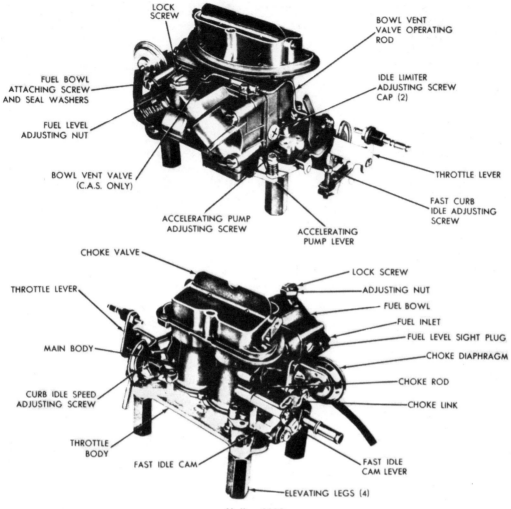

LOCK SCREW

BOWL VENT VALVE OPERATING ROD

FUEL BOWL ATTACHING SCREW AND SEAL WASHERS

IDLE LIMITER ADJUSTING SCREW CAP (2)

FUEL LEVEL ADJUSTING NUT

THROTTLE LEVER

BOWL VENT VALVE (C.A.S. ONLY)

FAST CURB IDLE ADJUSTING SCREW

ACCELERATING PUMP ADJUSTING SCREW

ACCELERATING PUMP LEVER

CHOKE VALVE

LOCK SCREW

THROTTLE LEVER

ADJUSTING NUT

FUEL BOWL

FUEL INLET

MAIN BODY

FUEL LEVEL SIGHT PLUG

CHOKE DIAPHRAGM

CURB IDLE SPEED ADJUSTING SCREW

CHOKE ROD

CHOKE LINK

THROTTLE BODY

FAST IDLE CAM

FAST IDLE CAM LEVER

ELEVATING LEGS (4)

Holley 2300

bores for wear or an out-of-round condition. Damage or wear to the throttle arm, shaft, or shaft bore will often require replacement of the throttle body. These parts require a close tolerance of fit; wear may allow air leakage, which could affect starting and idling.

NOTE: *Throttle shafts and bushings are not included in overhaul kits. They can be purchased separately.*

4. Inspect the idle mixture adjusting needles for burrs or grooves. Any such condition requires replacement of the needle, since you will not be able to obtain a satisfactory idle.

5. Test the accelerator pump check valves. They should pass air one way but not the other. Test for proper seating by blowing and sucking on the valve. Replace the valve if necessary. If the valve

is satisfactory, wash the valve again to remove breath moisture.

6. Check the bowl cover for warped surfaces with a straightedge.

7. Closely inspect the valves and seats for wear and damage, replacing as necessary.

8. After the carburetor is assembled, check the choke valve for freedom of operation.

Carburetor overhaul kits are recommended for each overhaul. These kits contain all gaskets and new parts to replace those that deteriorate most rapidly. Failure to replace all parts supplied with the kit (especially gaskets) can result in poor performance later.

Some carburetor manufacturers supply overhaul kits of three basic types: minor repair, major repair; and gasket kits. Basically, they contain the following:

Minor Repair Kits:
- All gaskets
- Float needle valve
- Volume control screw
- All diaphragms
- Spring for the pump diaphragm

Major Repair Kits:
- All jets and gaskets
- All diaphragms
- Float needle valve
- Volume control screw
- Pump ball valve
- Float
- Complete intermediate rod
- Intermediate pump lever

- Some cover hold-down screws and washers

Gasket Kits:
- All gaskets

After cleaning and checking all components, reassemble the carburetor, using new parts and referring to the exploded view. When reassembling, make sure that all screws and jets are tight in their seats, but do not overtighten, as the tips will be distorted. Tighten all screw gradually, in rotation. Do not tighten needle valves into their seats; uneven jetting will result. Always use new gaskets. Be sure to adjust the float level when reassembling.

Carter AVS

	1968	1969	1970	1971
Fast Idle Speed (rpm)	Auto. 1400 Man. 1600	1700	1800	①
Float Setting (in.)	5/16	340 engine 7/32 383 engine 5/16	7/32	7/32
Float Drop (in.)	1/2	1/2	1/2	1/2
Choke Unloader (in.)	1/4	1/4	1/4	1/4
Vacuum Kick (in.)	Auto. 0.15 Man. 0.2	Auto. 0.11 Man. 0.07	0.16	0.09
Idle Mixture Screws (turns out)	1–2	1–2	——	——

Auto.—Automatic transmission
Man.—Manual transmission
① 383 engine, auto.—1700
   440 engine, auto.—1800
   440 engine, man.—2100

Carter/Ball and Ball BBD

	1968	1969	1970	1971	1972	1973	1974–75	1976
Fast Idle Speed (rpm)	1600	1600	1700	Auto. 1700 Man. 1900	1700②	1700	1500	1200
Float Setting (in.)	5/16	5/16	5/16	5/16	1/4	1/4	1/4	1/4
Choke Unloader (in.)	1/4	1/4	1/4	1/4	1/4	1/4	21/64	5/16
Vacuum Kick	Auto. 0.18 Man. 0.23	5/16	0.16	Auto. 0.14 Man. 0.16	0.15	0.15①	0.11	0.07

Carter/Ball and Ball BBD (cont.)

	1968	1969	1970	1971	1972	1973	1974–75	1976
Idle Speed Screws (turns out)	1½	1½	1½	1½	—	—	—	—

Auto.—Automatic transmission
Man.—Manual transmission
① 318 engine, Auto., Non-California, 0.13 in.
② California cars—Auto.—1800 rpm

Holley 1920

	1968–69	1970	1971
Fast Idle Speed (rpm)	Auto. 1800 Man. 1600	Auto. 700 Man. 650	Auto. 1900 Man. 1600
Float Setting	Use gauge	Use gauge	Use gauge
Choke Unloader (in.)	⁹⁄₃₂	⁹⁄₃₂	⁹⁄₃₂
Vacuum Kick	Auto. 0.07 Man. 0.10	0.10	0.10
Idle Mixture Screws (turns out)	2	—	—

Auto.—Automatic transmission
Man.—Manual transmission

Holley 2210

	1972	1973
Fast Idle Speed (rpm)	1900	1800
Dry Float Setting (in.)	0.18	0.18
Choke Unloader (in.)	0.17	0.17
Vacuum Kick (in.)	0.11	0.15

Holley 2245

	1974–75	1976
Fast Idle Cam	0.110	0.110
Fast Idle Speed (rpm)	①	1600
Float Setting (in.)	0.180	0.180
Choke Unloader (in.)	0.170	0.170
Vacuum Kick (in.)	0.150	0.150

① 360V8—1800 rpm
400V8—1600 rpm

Holley 2300

	1970	1971	1972
Fast Idle Speed (rpm)	2200	1800	1800
Float Setting	Center float in bowl	Center float in bowl	Center float in bowl
Choke Unloader (in.)	$\frac{5}{32}$	$\frac{5}{32}$	0.15
Vacuum Kick (in.)	0.07	0.07	0.07

Holley 4160

	1968	1969	1970	1971	1972
Fast Idle Speed (rpm)	700①	1400①	1600	1700	Auto. 1600③
Dry Float Setting (in.)	Primary $\frac{7}{64}$ Secondary $\frac{15}{64}$	$\frac{15}{64}$ $\frac{17}{64}$	$\frac{15}{64}$ $\frac{17}{64}$	$\frac{15}{64}$ $\frac{17}{64}$	0.110 0.204
Wet Float Setting (in.)	Primary $\frac{9}{16}$ Secondary $\frac{13}{16}$	$\frac{9}{16}$ $\frac{13}{16}$	$\frac{9}{16}$ $\frac{13}{16}$	$\frac{9}{16}$ $\frac{13}{16}$	$\frac{9}{16}$ $\frac{13}{16}$
Choke Unloader (in.)	$\frac{5}{32}$	$\frac{5}{32}$	0.15	0.15	0.15
Vacuum Kick	0.06	0.08	0.14	②	Auto. 0.08
Idle Mixture Screws (turns out)	1–1¼	1–1¼	—	—	—

① On step no. 5
② Fresh air, auto.—0.14
   Fresh air, man.—0.08
   Heated air, auto.—0.08
   Heated air, man.—0.14
③ California cars, auto.—1800
Auto.—Automatic transmission
Man.—Manual transmission

Rochester 2GV

	1971
Fast Idle Speed (rpm)	1800
Float Setting (in.)	$\frac{21}{32}$
Float Drop (in.)	1¾
Choke Unloader (in.)	0.136
Vacuum Kick (in.)	0.096

Thermo-Quad® (Carter)

	1971	1972	1973	1974–75	1976
Fast Idle Speed (rpm)	900	Auto. 750 ② Man. 900	1700①	③	④
Float Setting (in.)	1	1	1¹⁄₁₆	1	29⁄32
Choke Unloader (in.)	0.190	0.190	0.190	0.310	0.310
Vacuum Kick (in.)	0.110	0.140	0.160	0.160	0.100

Auto.—Automatic transmission
① 400 engine, Auto.—1800
② California cars, Auto.—750
③ 360 engine—1800
    400 engine—2000
    400 HP engine—1800
    440 engine—1700

④ 360—1700 rpm
    400 Calif—1600 rpm
    400 49 states—1800 rpm
    440—1600 rpm

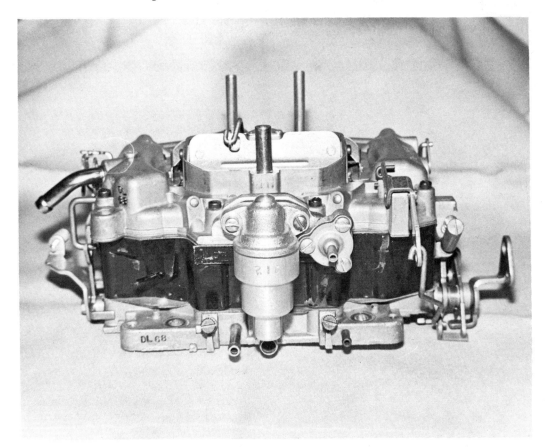

The large device at the center of the carburetor is an altitude compensator; the small nozzle at the right is the idle enricher—1975–76 Carter Thermo-Quad®

Chassis Electrical

HEATER

Blower

REMOVAL AND INSTALLATION

With or Without Air Conditioning

The blower motor is mounted to the engine side housing under the right front fender, between the inner fender shield and the fender. The inner fender shield must be removed to service the blower motor.

1. Raise the hood and remove all brackets and clips that attach to the inner fender shield under the hood.

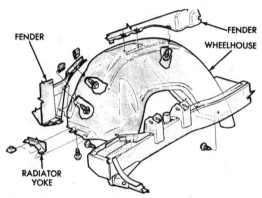

Inner fender shield attaching points

2. Raise the car on a hoist and remove the right front tire and wheel assembly.

3. From under the fender, remove the bolts that attach the inner fender shield to the fender.

4. Remove the fender shield from the vehicle.

5. Disconnect the blower motor wiring at the multiple connector.

6. Remove the nuts that attach the blower motor to the heater housing and remove the blower motor.

Core

REMOVAL AND INSTALLATION

Without Air Conditioning

NOTE: *This is the removal procedure for the heater housing that attaches to the passenger compartment side of the firewall. Do not remove the part of the housing that attaches to the engine side of the firewall.*

1. Disconnect the battery and drain the radiator.

2. Disconnect the heater hoses at the dash panel. Plug the hose fitting on the heater to prevent spilling coolant on trim.

3. Slide the front seat back to allow room.

4. Disconnect the radio antenna.

5. Disconnect the electrical conductors from the blower motor resistor block on the face of the housing.

6. Remove the vacuum hoses from the trunk lock if so equipped.

7. Remove the control cables from the defroster door crank and heat shut off door crank.

8. Remove the bottom retaining nut from the support bracket and swing the bracket up and out of the way.

9. In the engine compartment remove the four retaining nuts from the studs on the engine side housing.

10. Remove the locating bolt from the bottom center of the passenger side housing.

11. Roll or tip the housing out from under the instrument panel.

12. Remove the temperature control cable retaining clip and the cable from the heat shut off door crank.

13. From inside the heater assembly, remove the two retaining nuts from the right-side of the heater core.

14. Remove the four heater core attaching screws from the outside of the heater housing.

15. Remove the heater core locating metal screw from the top of the heater housing.

16. Carefully pull the heater core from the heater housing.

17. Installation is the reverse of removal.

With Air Conditioning—1968–71

The air conditioning heater core is located in the front cover of the passenger side housing. To remove only the heater core, the air conditioning system need not be discharged.

1. Disconnect the battery and drain the cooling system. Disconnect the heater hoses and remove the air cleaner.

2. Plug the heater core tubes to prevent coolant loss when the core is removed.

3. Take off the steering column cover and remove the left spot cooler duct.

4. At the linkage on the left side of the housing, disconnect the two actuator rods. Remove the two cover retaining screws.

5. Remove the screws securing the

heat duct in position and remove the duct. With the duct removed, the screws in the bottom of the front cover lip will be exposed. Remove them.

6. Remove the glove box. Remove the center spot cooler duct, the air distribution housing, and the right spot cooler duct.

7. Working in the glove box opening, remove the top retaining screws and the screw from the right side of the housing.

8. At the resistor block, disconnect all electrical connections. Disconnect the vacuum hoses from the recirculating housing actuator.

9. Take off the nut at the housing end of the cover support bracket. Swing the bracket upward and carefully roll the front cover and heater core outward. Remove it from under the instrument panel.

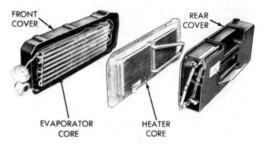

FRONT COVER   REAR COVER   EVAPORATOR CORE   HEATER CORE

Heater core and evaporator—1968–71

10. To begin installation, replace the heater core in the front cover. Position the core and cover on the evaporator housing. While holding the front cover in position, swing the support bracket downward over the stud on the front cover face. Install its retaining nut.

11. Working in the glove box opening, replace the top housing and right-side screws.

12. Working under the instrument panel, install the screws that retain the housing in position.

13. Replace the heat distribution duct to the housing bottom and connect the actuator rods.

14. Connect all the vacuum hoses to their actuators; install all electrical connections to the resistor block.

15. Through the glove box opening, install the air distribution housing, the center spot cooler duct, and the right spot cooler duct.

16. Replace the steering column cover

and the left spot cooler duct. Replace the glove box assembly.

17. From this point, reverse the removal procedure. Be sure to fill cooling system with proper type and amount of antifreeze.

With Air Conditioning—1972-73

To remove only the heater core, it is not necessary to discharge the air conditioning system. The heater core is positioned in the rear housing of the passenger-side unit.

1. Disconnect the battery and drain the cooling system. Remove the air cleaner and disconnect the heater hoses. Plug the heater core tubes to prevent coolant loss when the core is removed.

2. Remove the steering column cover and remove the left spot cooler duct.

3. On the left side of the housing, remove the linkage shield and disconnect the actuator rods. Remove the screws from the housing left side.

4. Remove the screws holding the heat distribution duct and remove it. With duct removed, the screws in the bottom lip of the rear housing will become visible. Remove them.

5. Remove the glove box. In addition, remove the right spot cooler duct, the air distribution housing, and the center outlet duct.

6. Working in the glove box opening, remove the top retaining screws and the right-side housing screws. If the vehicle is Auto-Temp equipped, remove the aspirator tube from the clip first and then remove the amplifier and master compressor switches. Now remove the right-side housing screws.

7. Disconnect all electrical connections at the resistor block. On Auto-Temp equipped vehicles, remove the wires from the two plastic straps and the metal clip.

8. Remove the nut from the housing end of the support bracket. Swing the bracket upward and out of the way. Carefully roll the housing out from under the instrument panel. The heater core may be removed by pulling it out from the top. Cut the adhesive along the bottom and sides with a knife to ease removal.

9. To begin installation, scrape all remaining sealer from the heater core flange and fit a new seal. Position the heater core in the rear housing and secure with a screw at either end. Place the front housing in position; hold the rear housing in place and swing the support bracket down. Secure it in position with its retaining nut.

10. Working in the glove box opening, install the top two housing screws and the screws at the right side of the rear housing.

11. From beneath the instrument panel, install the screws along the housing bottom and the screws at the left side of the rear housing. It is not necessary to reinstall the linkage shield.

12. Replace the heat distribution duct to the housing bottom.

13. Connect the actuator rods.

14. Working in the glove box opening, connect the resistor block wires. Tighten the support bracket nuts. On Auto-Temp equipped vehicles, fasten the wires with the plastic straps and metal clip. Install the aspirator tube in the clip.

15. Replace the center outlet duct, and air distribution housing, and the right spot cooler duct.

16. Install the steering column cover, the left spot cooler duct, and the glove box assembly. On Auto-Temp equipped vehicles, install the amplifier and the master and compressor switches.

17. From this point, reverse the removal procedure. Be sure to fill the cooling system with the proper amount and type of anti-freeze.

With Air Conditioning—1974-76

In order to remove the heater core on 1974–76 models with air conditioning, it is necessary to discharge the air conditioning coolant lines. Because discharging and recharging these lines is a hazardous operation which also requires the use of special equipment, it is recommended that this procedure is done by qualified service personnel.

RADIO

REMOVAL AND INSTALLATION

CAUTION: *Never operate the radio without a speaker, damage to the output transistors will result. If the speaker must be replaced, use a replacement of the correct impedance (ohms) for the radio, or else the output transistors will be damaged and will require replacement.*

1968 Fury and VIP

1. Remove the instrument cluster bezel. (See Instrument Cluster Removal in this chapter).
2. From under the panel, loosen the radio support bracket nut at the upper end.
3. Disconnect the feed wires, speaker wires, and antenna cable at the radio.
4. From the front of the instrument panel, remove three radio mounting screws and lift the radio out of the panel.
5. Reverse the procedure to install.

1969–73 Fury

1. Disconnect the battery.
2. Remove nine lamp panel mounting screws, lower the lamp panel assembly slightly, disconnect the lamp harness from the main harness, and remove the lamp panel from the instrument panel.
3. Remove the steering column cover.
4. Remove the radio trim bezel mounting screws and bezel.
5. Remove the center lower air conditioner duct, if so equipped.
6. Disconnect all electrical leads and the antenna lead at radio.
7. Remove the radio support mounting bracket.
8. Remove the two radio mounting bolts.
9. Move the radio down through the bottom of instrument panel carefully to avoid damage to vacuum hoses and electrical leads. Reverse the procedure to install.

1974 Fury and 1975–76 Gran Fury

1. Remove the screws securing the instrument cluster bezel and then remove the bezel (see "Instrument Cluster," below).
2. Remove the speedometer bezel.
3. On monaural radios, remove the light assembly from the front of the radio.
4. Lift the back edge of the instrument panel trim cover to release the

mounting clips. Lift the cover rearward and then up, using care to make sure that the clips clear the trim pad surface.
5. Unfasten the screws which secure the radio to the panel.
6. Disconnect the antenna lead and remove the radio mounting bracket nut, working through the access hole in the top of the instrument panel.
7. On monaural radios, remove the speaker lead from the speaker; on stereo radios, disconnect the speaker leads at the radio.
8. Withdraw the radio from the panel and disconnect its power lead.

Installation is performed in the reverse order of removal. Use care not to damage the speaker cover with the upper cover installation clips.

WINDSHIELD WIPERS

Blade and Arm
REPLACEMENT

To replace the blade assembly or the blade element, proceed as follows:
1. Turn the ignition switch on, turn the wiper switch on, and then turn the *ignition* switch off, once the blades are conveniently positioned on the glass.
2. Raise the blade assembly off the glass by lifting the wiper arm up.
3. Depress the release lever (use a screwdriver if necessary) where the blade assembly is attached to the arm and separate the assembly from the arm.
4. Push the red (or black) button on end bridge and separate it from the center bridge.
5. Slide the rubber element out of the end bridges.
6. Install the new rubber refill element, being careful to engage it in all four end bridge claws securely.
7. Install the end bridge on the center bridge and the center bridge on the arm, making sure that they are securely locked in place.

To remove the arm assembly from the pivots, proceed as follows:
1. Open the hood on models with concealed wipers, to gain access to the wiper arms.

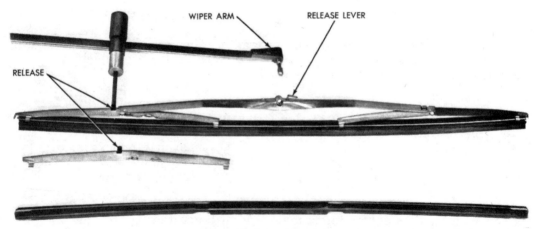

WIPER ARM

RELEASE LEVER

RELEASE

Wiper blade replacement

2. Use an 0.090 in. pin or drill to trip the wiper arm release on models made prior to 1971. On 1971–76 models, *lift* the release latch with your finger.

3. Slide the arm upward, off the pivot, using a rocking motion.

CAUTION: *Do not use a screwdriver to pry the arm off the pivot; it may become distorted. Do not push or bend the spring clip in the base of the arm in an attempt to remove the arm.*

4. Install the arms by sliding them over the pivots so that the blades contact the concealed wiper blade stops or so that the blades are ¼–2 in. above the windshield molding on non-concealed wiper models, with the motor in the park position.

5. Operate the wipers at low speed; the blades should not contact the wind-

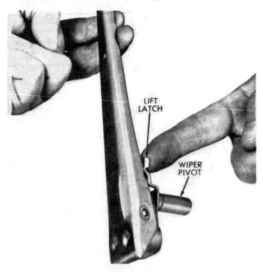

LIFT LATCH

WIPER PIVOT

Removing the wiper arm—1971–76

shield moldings at either end of their sweep. If they do, remove the arms and reposition them until correct clearance is obtained.

NOTE: *Be sure that the blades are concealed when parked on models having concealed wipers.*

Motor

REMOVAL AND INTALLATION

1968–70

1. Remove the windshield wiper arm and blade assemblies. On 1969–70 models insert a 0.090 in. pin in the hole in the base of the wiper arm to release the assemblies from the pivots.

2. Remove the windshield lower moulding.

3. Remove the cowl grille.

4. Remove the nut that attaches the wiper link to the wiper motor drive pin or crank and disconnect the link from the motor.

5. Disconnect the wiper motor wiring at the multiple connector.

6. Remove the nuts that attach the wiper motor to the cowl panel and remove the motor through the cowl grille opening.

7. Installation is the reverse of removal.

1971–76

1. Disconnect the negative battery cable.

2. Lift the latch on each wiper arm and remove the arms and blades as an assembly.

3. Remove the cowl screen.

4. Remove the drive crank retaining nut and drive crank.

5. Disconnect the lead wires from the wiper motor.

6. Remove the three wiper motor mounting bolts and remove the motor from the vehicle.

7. Reverse the above procedure to install. When installing the wiper arms and blades, make sure the wiper motor is in the park position.

Linkage

REMOVAL AND INSTALLATION

1968–69

1. Remove the wiper arm and blade assemblies.

2. Remove the windshield lower moulding.

3. Remove the cowl grille.

4. To remove the right pivot, disconnect the connecting link from the pivot, remove the bolts that attach the pivot to the cowl and remove the pivot.

5. To remove the left pivot, disconnect the connecting link from the right pivot. Disconnect the drive crank from the wiper motor. Remove the bolts that attach the left pivot to the cowl and remove the pivot and links through the cowl opening.

6. Installation is the reverse of removal.

1970

1. Remove the wiper arm and blade assemblies.

2. Remove the cowl screen.

3. Remove the crank arm nut and crank arm from the wiper motor.

4. Remove the bolts that attach the right and left pivots to the body.

5. Remove the pivots and linkage from the vehicle as an assembly through the cowl opening.

6. Installation is the reverse of removal.

1971–76

1. Remove the arm and blade assemblies from the wiper pivots.

2. Remove the cowl screen.

3. Remove the crank arm retaining nut and crank arm from the wiper motor.

4. Remove the bolts that attach the right and left pivots to the body of the vehicle.

5. Remove the links and pivots as an assembly through the cowl opening.

6. Reverse the above procedure to install.

INSTRUMENT CLUSTER

Removal and Installation

1968

1. Disconnect the battery.

2. Remove the eight instrument cluster light panel retaining screws, remove panel and rest panel on top of trim pad. It is not necessary to disconnect the wiring.

3. Remove the heater or air conditioning control knobs, and the clock reset knob.

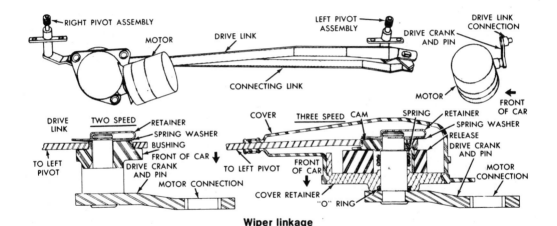

Wiper linkage

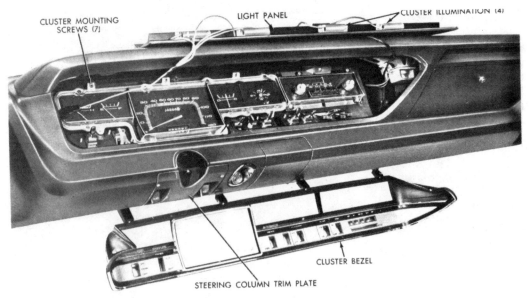

CLUSTER MOUNTING SCREWS (7)   LIGHT PANEL   CLUSTER ILLUMINATION (4)

CLUSTER BEZEL

STEERING COLUMN TRIM PLATE

1968 instrument cluster removal

4. Remove the six bezel retaining screws and remove bezel.

5. Remove the four screws from steering column cover and drop the cover with vent controls attached.

6. In vehicles with automatic transmission and column shift, remove the gear selector link nut, spring washer and bolt.

7. Remove the four stereo speaker grille screws and place the speaker on top of the instrument panel.

8. Disconnect the speedometer cable, and remove the five cluster mounting screws. Raise the cluster slightly, roll its upper edge out and disconnect ammeter leads.

9. With the cluster face down disconnect fuel and temperature gauge wires and high beam, oil pressure, and turn signal light sockets.

10. Remove the cluster. Reverse the procedure to install.

1969-71

1. Disconnect the negative battery cable.

2. Remove the lamp panel.

3. Remove the steering column cover.

4. Remove the radio cover bezel.

5. Remove the instrument panel left trim bezel and spot cooler (if so equipped).

6. If equipped with air conditioning,

remove the left-side duct and center air conditioning connector.

7. Working under the dash, disconnect all electrical leads from the cluster and disconnect the speedometer cable from the speedometer head.

8. Remove the gear shift indicator pointer from the steering column.

9. Tape steering column to protect the paint finish.

10. Remove the three steering column upper clamp nuts and the three bolts from the steering column lower support at the floor.

11. Lower the steering column and allow it to rest on the front seat cushion.

12. Remove the eight screws that attach the instrument cluster to the instrument panel. Roll the cluster forward and disconnect any electrical connections still attached to the cluster as they become accessible.

13. Remove the cluster from the vehicle.

14. Reverse above procedure to install.

1972-73

1. Disconnect the negative battery cable.

2. Remove the instrument panel lamp housing attaching screws.

3. Pull the lamp housing down and disconnect the lead wire from it, then

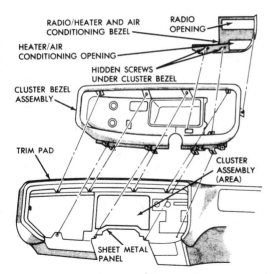

RADIO/HEATER AND AIR CONDITIONING BEZEL

RADIO OPENING

HEATER/AIR CONDITIONING OPENING

HIDDEN SCREWS UNDER CLUSTER BEZEL

CLUSTER BEZEL ASSEMBLY

TRIM PAD

CLUSTER ASSEMBLY (AREA)

SHEET METAL PANEL

1972–73 instrument cluster removal (1969–71 similar)

remove the lamp housing from the instrument panel.

4. Remove the hood release handle from the steering column cover.

5. Remove the steering column cover attaching screws.

6. If equipped with a radio fader switch, lower the column cover and disconnect the lead wires from the fader switch.

7. Disengage the vent control cable from the steering column cover by pulling the vent handle out slightly and guiding the shaft up through the slot in the steering column cover. Remove the vent control assembly through the hole in the steering column cover.

8. If equipped with air conditioning, disconnect the left spot cooler duct from the left spot cooler housing and the center air conditioner connector from the airconditioner housing.

9. Lower the steering column.

10. Push down on the flat tab on the speedometer cable-to-speedometer housing connector and disconnect the cable from the housing.

11. Remove the radio cover bezel.

12. Remove the 10 instrument cluster retaining screws (two hidden under radio bezel).

13. Roll the instrument cluster forward and disconnect the electrical connectors from the cluster as they become accessible.

14. Remove the cluster from the dash.

15. Reverse the above procedure to install.

1974–76

1. Disconnect the negative (−) battery cable.

2. Place the transmission gear selector in the "1" range.

3. Remove the ash tray from its housing and remove the cigarette lighter.

4. Center the windshield wiper/washer, rear window defroster, heater/

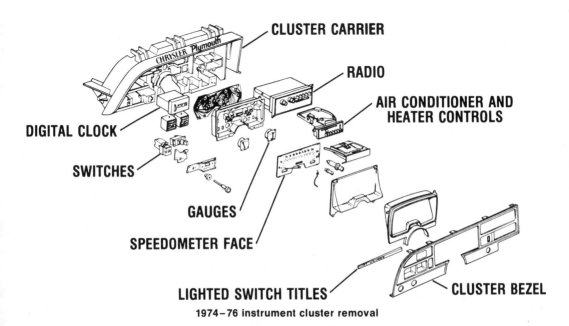

CLUSTER CARRIER

RADIO

AIR CONDITIONER AND HEATER CONTROLS

DIGITAL CLOCK

SWITCHES

GAUGES

SPEEDOMETER FACE

LIGHTED SWITCH TITLES

CLUSTER BEZEL

1974–76 instrument cluster removal

AC fan, and heater/AC temperature control switches in their openings.

5. Unfasten the cluster bezel securing screws from the lower edge of the cluster. Remove the bezel by tipping its top outward to release its clips and then disengaging its locktabs.

6. Withdraw the nylon pins which secure the speedometer bezel with pliers, and remove the bezel.

7. Remove the transmission gear indicator needle.

8. Withdraw the pins which secure the cluster lens with pliers and remove the lens.

9. Lift the back edge of the upper instrument panel trim cover to release its mounting clips. Lift the cover rearward and then up, using care that the clips clear the trim pad surface.

10. Working through the hole in the top of the instrument panel, unfasten the speedometer cable and disconnect the two instrument cluster multiconnectors.

11. Unfasten the screws which secure the cluster housing to the carrier.

12. Carefully lift out the cluster assembly.

13. Installation is the reverse of removal.

Ignition Switch

These procedures cover presteering column-mounted switches and lock cylinders; for 1970–76 steering column-mounted switch removal procedures, see Chapter 8.

REMOVAL AND INSTALLATION

1968

1. Disconnect the push-on connector from the rear of the ignition switch, directly behind the ignition lock cylinder.

2. Remove the bezel nut that attaches the ignition switch to the rear of the instrument panel.

3. Remove the ignition switch.

4. Position a new ignition switch under the instrument panel and install and tighten the bezel nut.

5. Connect the wiring to the rear of the switch.

Ignition Lock Cylinder
REMOVAL AND INSTALLATION
1968–69

1. Insert the ignition key into the lock cylinder.

2. Insert a piece of stiff wire into the small hole in the front face of the cylinder and apply pressure to the wire.

3. Turn the ignition key counterclockwise toward the "ACC" position.

4. Pull the lock cylinder and key from the instrument panel.

5. Insert a new lock cylinder into the instrument panel and it will lock itself in place.

SEAT BELT SYSTEMS

Light and Buzzer—1972–73

All Chrysler Corporation cars built after January 1, 1972 have front seat belts with a reminder light and buzzer system. The warning system consists of a buzzer and light, two lap belt retractor switches, a front passenger weight-sensing switch and a relay.

With only the driver in the front seat, the system will give warning when the ignition switch is on, the automatic transmission is in any gear or the parking brake is released on manual transmission models and the driver's seat belt has not been pulled out of the retractor at least ten inches.

With a passenger in the right front seat, the system will operate as for the driver alone unless the passenger's seat belt has also been extended.

Seat Belt/Starter Interlock—1974–75

The interlock system prevents the starter from being operated until all outboard front seat occupants have fastened their seat belts. An interlock control requires that the seat belts be fastened in all outboard front seats each time the engine is started.

For the convenience of the serviceman, there is an underhood bypass switch for the system. This enables him to start the engine once, without having to sit on the seat and fasten the seat belt.

The earlier light and buzzer reminder

TORQUEFLITE TRANSMISSION MODELS

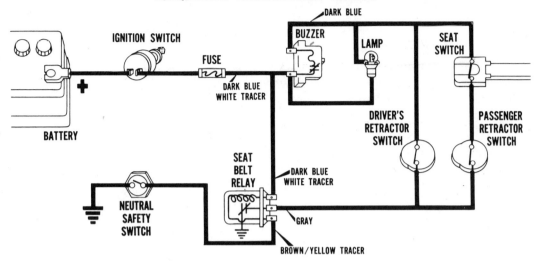

Seat belt warning system—1972–73

system is retained, with coverage for the center front seat position added.

COMPONENT LOCATIONS

The underhood by-pass switch is located near the right-hand hood hinge plate on all Plymouth models.

The remaining major components are situated in the following locations: Buzzer—at the left-side of the brake support bracket on 1974 full size models. Buzzer Light Fuse—fuse #3 on 1974-75 full size models. Interlock Unit—above the buzzer on 1974–75 full sized models.

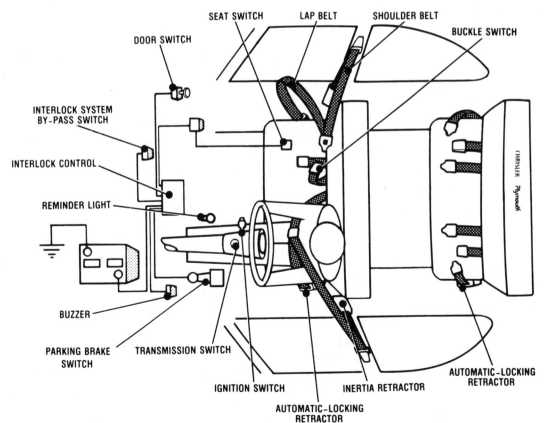

Seat belt interlock system—1974–75

Interlock System Fuse—fuse #5 on 1974–75 full sized cars.

CHECKOUT PROCEDURE

Chrysler makes available to its dealers a special electronic interlock system checkout device. However, for those who encounter a no-start condition attributable to the starter interlock system, the following numbered check out procedure may be used as a substitute. Necessary tools are a 12 volt test light and a test light jumper wire and probe. Remember when checking out a circuit, test only at the wire end of the interlock connectors.

Always begin the test procedure with the transmission in Neutral (N) on automatic transmission equipped cars, and with the parking brake applied on manual transmission equipped cars. Always place the ignition key in the "OFF" or "LOCK" position. The system must be activated by placing an adult weighing 150 lbs or an equivalent weight momentarily on the seat before starting the test procedure.

NOTE: *The following procedure should be used in conjunction with the "Interlock Start Circuit Diagnosis Chart"*

Underhood Switch Tests

1. Sit in the driver's seat with all doors closed, but do not buckle the belt. Turn the ignition key to "START". Engine should not crank. Light and buzzer should go on while the key is in "START". Light and buzzer should go "OFF" when the key is in "RUN".

2. Turn the ignition key to "OFF" or "LOCK". Open the hood and depress, then release the underhood switch. Sit in the driver's seat, but do not buckle the belt. Turn the ignition key to "START". The engine should crank, and the light and buzzer should go on while key is in "START" position.

3. Leave ignition on for 30 seconds, then turn it off and leave the ignition off for 2.5 minutes. Get out of the driver's seat and reach in car and buckle driver's seat belt. Sit in the driver's seat, close all of the doors again and turn the ignition

key to "START". The engine should not crank, and the light and buzzer should go on while the key is in the "START" position.

Interlock and Sensing Switches Test

4. With the driver in the seat and pre-buckled as in Step 3 above, leave the ignition "ON" and move the gear selector to any forward or reverse gear. Light and buzzer should go on and stay on.

5. Unbuckle the driver's belt and light and buzzer should stay on. Kneel on the seat and buckle belt across empty seat. In this way weight applied to driver's seat before belt is buckled, light and buzzer should go off.

6. Get out of the car and close the door. Go to the right-side of the car and open the door. Reach in and buckle the right passenger's seat belt. Sit on the seat and close the door and the light and the buzzer should go on.

7. Unbuckle the right passenger's belt and the light and the buzzer should go on.

8. Buckle the right passenger's belt and the light and the buzzer should turn off.

9. Reach over and buckle the center passenger's belt. Then unbuckle the right passengers belt and move to center seating position. The light and the buzzer should go on.

10. Unbuckle the center seat belt and the light and the buzzer should go on.

11. Buckle the center seat belt and the light and the buzzer should turn off within five seconds.

12. Unbuckle the center seat belt and move to the driver's seat, then place the automatic transmission gear selector in Neutral (N) or Park (P). Turn the ignition key to "START" and the engine should not crank. Unbuckle and rebuckle the driver's belt. Engine should crank and light should be off.

13. Turn ignition to "OFF" and unbuckle the driver's belt. Next, turn the ignition key to start. Engine should crank. The light and buzzer should operate while the key is in the "START" position.

14. Turn the ignition key off and move to the center or right seat for at

Interlock Start Circuit Diagnosis Chart

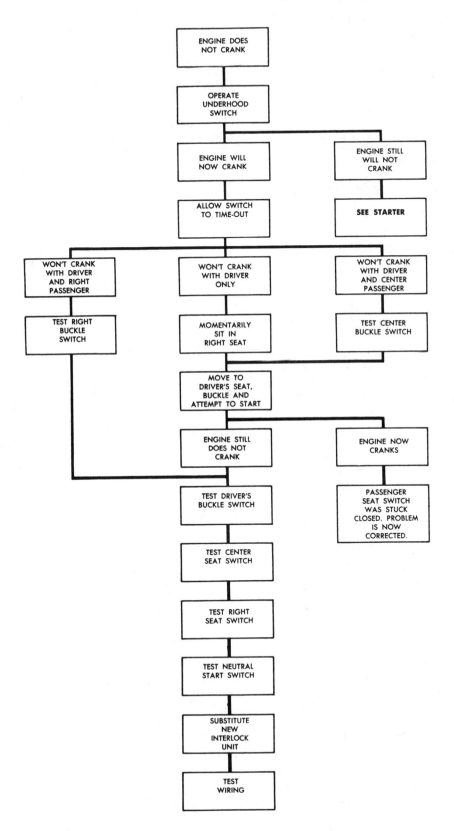

least five seconds. Return to the driver's seat and turn the ignition switch to "START" and the engine should crank. The light and the buzzer should operate while the key is in the "START" position.

HEADLIGHTS

REMOVAL AND INSTALLATION

1. On models with concealed headlights, turn the ignition switch to the "ON" position and then turn the headlights on. Next, turn the headlights off, after turning the ignition switch off; this will allow the headlight doors to remain opened with the headlights off.

2. Unfasten the screws which secure the headlight bezel and remove the bezel.

3. Remove the three headlight retaining ring screws and remove the ring while supporting the sealed beam unit with your hand.

CAUTION: *Do not disturb the two headlight aiming screws.*

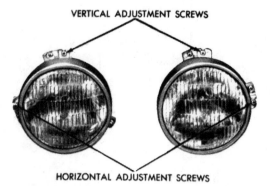

VERTICAL ADJUSTMENT SCREWS

HORIZONTAL ADJUSTMENT SCREWS

Headlight aiming screw location—do not remove to replace sealed beam unit

4. Remove the sealed beam unit from the car and pull the electrical connector straight off the back.

5. Installation is performed in the reverse order of removal.

CAUTION: *Do not interchange the inner and outer headlight sealed beam units on models with dual headlights.*

Light Bulbs

1968

Air conditioning indicator	FL
Ash receiver	FL
Back-up lights	1073
Brake system warning light	57
Dome and/or "C" pillar light	1004
Door, pocket panel and/or reading light	90
Fender mounted turn signal indicator	330M
Gear selector indicator	FL
Gear selector with console	57
Glove compartment	1891
High beam indicator	57
Instrument cluster illumination	1893
Ignition switch	1445
License light	67
Map light	1445
Oil pressure indicator	57
Park and turn signal	1034NA
Radio	FL
Sealed beam—hi beam (No. 1)	4001
Sealed beam—hi-lo beam (No. 2)	4002
Side Marker	1895
Tachometer	FL
Tachometer with console	1816
Tail, stop and turn signal	1034
Trunk and/or under hood light	1004
Turn signal indicator (panel)	57

1969

Air conditioner control and auto-temp.	**1893 (FL)
Ash receiver	**1445
Back-up lights	1156 (2)
Brake system warning indicator	57
Clock	**1893 (FL)
Cornering lights	1293 (2)
Dome lamp	551
Pocket panel lamp	90
Fender mounted turn signal indicator	330 (2)
Gear selector indicator	**1893 (FL)
Gear selector with console	** 57
Glove compartment	1891
High beam indicator	1892
Instrument cluster and speedometer illumination	**1893 (FL)
Ignition lamp	1445
License light	704
Map and courtesy lamp	90
Oil pressure indicator	57
Park and turn signal	1157 (2)
Radio	**1893 (FL)
Sealed beam—hi beam (No. 1)	4001
Sealed beam—hi-lo beam (No. 2)	4002
Stereo indicator	1445
Switch lighting	**1893 (FL)
Tachometer	**1893 (FL)
Tail, stop and turn signal	1157 (2)
Trunk and/or under hood lamp	1004
Turn signal indicator (panel)	57 (2)

Light Bulbs (cont.)

1969

Reverse 4-speed transmission indicator	53
Heat control	1893

1970

Air conditioner control and auto-temp.	1893
Ash receiver	1445
Back-up lights	1156
Brake system warning indicator	57
Clock	1893
Cornering lights	1293
Dome lamp	550
Pocket panel lamp	90
Fender mounted turn signal indicator	330
Gear selector indicator	1893
Gear selector with console	57
Glove compartment	1891
High beam indicator	1892
Instrument cluster and speedometer illumination	1893
Ignition lamp	1445
License light	67
Map and courtesy lamp	90
Oil pressure indicator	57
Park and turn signal	1157
Radio	1893
Sealed beam—hi beam (No. 1)	4001
Sealed beam—hi-lo beam (No. 2)	4002
Switch lighting	1893
Tachometer	1893
Tail, stop and turn signal	1157
Trunk and/or under hood lamp	1004
Turn signal indicator (panel)	57
Reverse 4-speed transmission indicator	53
Heat control	1893

1971

Air conditioner control and auto-temp.	*
Ash tray	1445
Back-up	1156
Brake system warning	57
Clock	*
Cornering	1293
Courtesy	90
Dome	211-2
Door	90
Door ajar	1892
Fender turn-signal	330-2
Front park and turn-signal	1157
Gear selector (column)	*
Gear selector (console)	*
Glove compartment	1891
High beam	1891
Instrument cluster	1893
Ignition	1445
License	1445
Low fuel	——
Map	——
Oil pressure	57

1971

Pocket panel	90
Radio	*
Reverse 4-speed transmission indicator	53
Sealed beam (No. 1)	4001
Sealed beam (No. 2)	4002
Side marker	1895
Stereo indicator	1445
Switches	1892
Tail, stop, and turn-signal	1157
Turn-signal indicator	57

1972–73

Air conditioner control	1445
Ash receiver	1445
Back up lights	1156
Brake system warning lights	57
Courtesy lamp	90
Dome and/or C pillar lamp	211-2
Door and pocket panel light	55
Door ajar indicator	57
Fasten belts indicator	57
Fender mounted turn signal	330
Front park and turn signal	1157
Fuel indicator	1892
Gear selector indicator (column)	*
Gear selector indicator (console)	57
Glove compartment	1891
High beam indicator	57
Instruments and speedometer	1816
Ignition lamp	1445
License light	67
Lock doors indicator	57
Oil pressure indicator	158
Sealed beam-hi beam (No. 1)	4001
Sealed beam-hi-lo beam (No. 2)	4000
Side marker	93
Switch lighting	1892
Tail, stop, turn signal	1157
Trunk, underhood lamp	1003-1004
Turn signal indicator (panel)	57

1974–76

Headlights	
Single	6014
Dual hi-beam	4001
Dual low-beam	4000
Side marker	168
Park and turn signal	1159NA
Tail/stop/turn signal	1157
Fender turn signal	168
License	
Sedan	168
Wagon	67
Back-up lights	1156
Speedometer	158
Clock	158
Heater/AC	363
Switches	
Headlight	1815
Wipe/Wash	1815
Rear defrost	1892
Rear unlock	1892

Light Bulbs (cont.)

1974–76

Cigarette lighter	158
Radio	168
Radio w/8-track	1893
Warning lights	158
Key light	1455
Glove box	1891
Dome lamp	211-2
Courtesy lamp	90
Trunk light	1003
Underhood light	1003
Seat belt warning	158
Door ajar	158
Hot temp indication	158
Washer fluid level	158

FL—Floodlighted
NA—Amber bulb
°—Included in instrument cluster lighting
°°—Headlamp rheostat dimming

FUSES AND FUSIBLE LINKS

Fuse Panel

LOCATION

1968

The fuse panel is located at the left-hand forward edge of the dashboard.

The turn signal flasher is mounted on a bracket which is attached to the ash tray bracket.

1969–73

The fuse panel is located under the dashboard to the left of the brake pedal. The turn signal flasher is located to the right of the steering column near the ash tray.

1974–76

The fuse panel is located to the left of the steering column, on the front edge of the dashboard, and swings down for easy fuse access once its locktab has been rotated 90°. All flashers and relays are mounted in a bank directly above the fuse panel.

Fuses and Circuit Breakers

1968

FUSES

Accessories	20 AMP
Cigar lighter (front) and dome light	20 AMP
Console (inline fuse)	20 AMP
Emergency flasher	20 AMP
Heater or air conditioner	20 AMP
Instrument lights	4 AMP

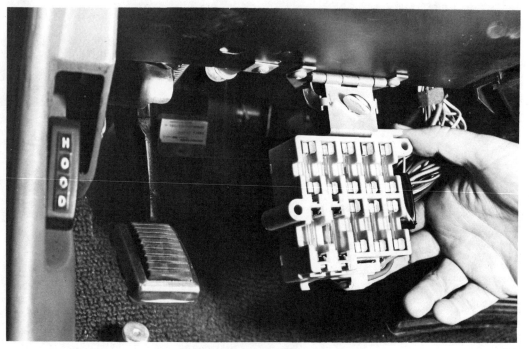

1974–76 swing-down fuse panel

Fuses and Circuit Breakers

1968

FUSES

Radio and back-up lamps	5 AMP
Tail stop	20 AMP

CIRCUIT BREAKERS

Cigar lighter—rear (behind left front cowl trim panel)	15 AMP
Convertible top (integral with top lift switch, behind left front cowl trim panel in Fury and VIP models)	30 AMP
Door locks (behind left front cowl trim panel)	15 AMP
Headlights (integral with headlight switch)	20 AMP
Power seats (behind left front cowl trim panel)	30 AMP
Power tailgate (instrument panel cluster behind ammeter)	30 AMP
Power windows (instrument panel cluster behind ammeter)	30 AMP
Windshield wipers (integral) with wiper switch)	7.5 AMP

1969–70

FUSES

Accessory	20 AMP
Console	20 AMP
Emergency flasher	20 AMP
Heater and air conditioner	20 AMP
Instrument lamps	3 AMP
Radio and back-up lamps	7.5 AMP°
Stop and dome lamps	20 AMP
Tail lamps and cigar lighter	20 AMP

°—1970—20

CIRCUIT BREAKERS

Convertible (on fuse block integral with switch)	30 AMP
Door locks (behind right front cowl trim panel)	15 AMP
Headlights (integral with headlamp switch)	20 AMP
Power seats (on fuse block)	30 AMP
Power tail gate (on fuse block)	30 AMP
Power windows (on fuse block)	30 AMP
Windshield wipers (integral with wiper switch)	7.5 AMP

1971

FUSES

Accessories	20 AMP
Console	20 AMP
Stoplights/dome light	20 AMP
Emergency flasher	20 AMP
Heater/air conditioner	20 AMP
Instrument lights	5 AMP
Miscellaneous	20 AMP
Radio/back-up lights	20 AMP
Tail lamps	20 AMP

CIRCUIT BREAKERS

Concealed headlamps (integral with relay, left end of instrument panel)	5 AMP
Door lock (on fuse block)	15 AMP
Front seat back latch (on fuse block)	15 AMP
Headlights (integral with switch)	20 AMP
Power seats (on fuse block)	30 AMP
Power tailgate (on fuse block)	30 AMP
Windshield wiper (integral w/wiper switch)	
2-speed	6 AMP
3-speed	7½ AMP

1972–73

FUSES

Power window, brake lamp	5 AMP
Horn	20 AMP
Hazard warning, stop lamp	20 AMP
Miscellaneous	20 AMP
Instrument lamps	5 AMP
Accessory	20 AMP
Heat, A/C	20 AMP
Radio, back-up lamps	20 AMP

INLINE FUSES

Console courtesy lamps	20 AMP
Headlamp delay relay	20 AMP
Heater and A/C blower motor when equipped with tilt column	20 AMP
Spotlight wiring	20 AMP
Electric deck lid solenoid/tailgate lock	20 AMP

CIRCUIT BREAKERS

Power door lock/front seat back latch	15 AMP
Power window, power seat, power tailgate window	20 AMP
Concealed headlamp relay	5 AMP

Fuses and Circuit Breakers   *1974–76*

1974–76

FUSES

Switch lamps, instrument cluster lights, tailgate lock	5 AMP
Clock, cigar lighter, tail lights, side marker lights, turn signal, parking lights, license lights	20 AMP
Trunk hazard, brake lights, seatbelt buzzer, dome, map, courtesy, glove box, seat belt lamps	20 AMP
Horns, horn relay	20 AMP
Power windows, warning lamps, trunk release, seat belt interlock, ignition feed	20 AMP
Radio	5 AMP
Accessories	20 AMP
Turn signal flasher, back-up lights voltage limiter, gauges	20 AMP

FUSES

Heater/AC blower motor/relay	20 AMP
AC high speed blower	30 AMP

INLINE FUSES

Automatic temperature control	1 AMP
Spot lights	10 AMP

CIRCUIT BREAKERS

Power door lights	15 AMP
Power windows, tailgate window, tailgate latch	30 AMP
Trailer lights, trailer (spare battery)	40 AMP

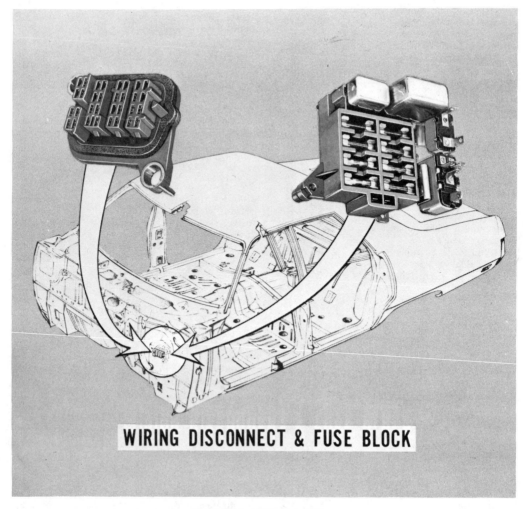

WIRING DISCONNECT & FUSE BLOCK

Pre-1974 fuse block location

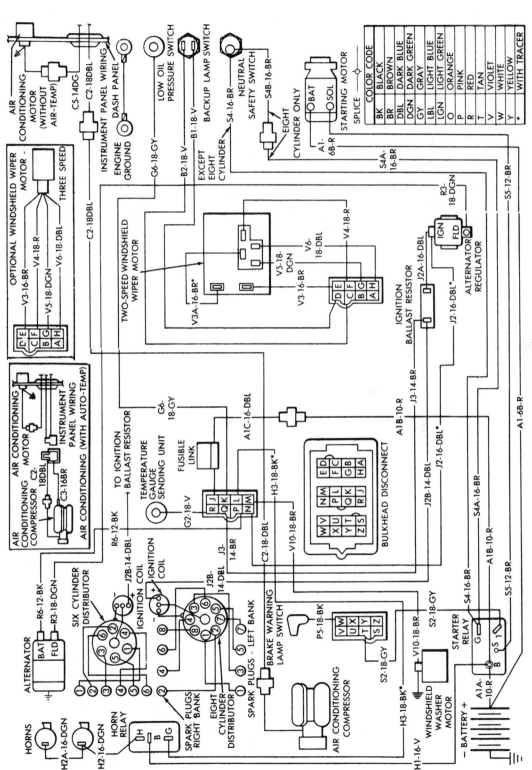

1968 engine compartment

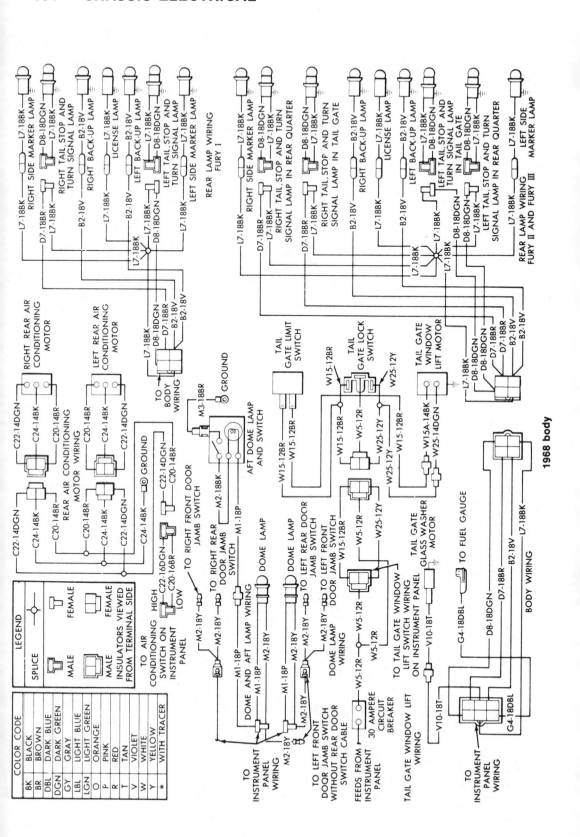

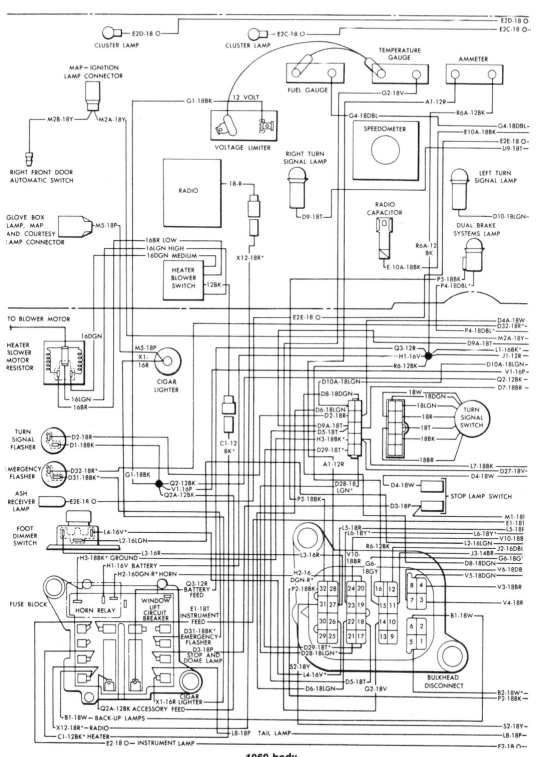

1969 body

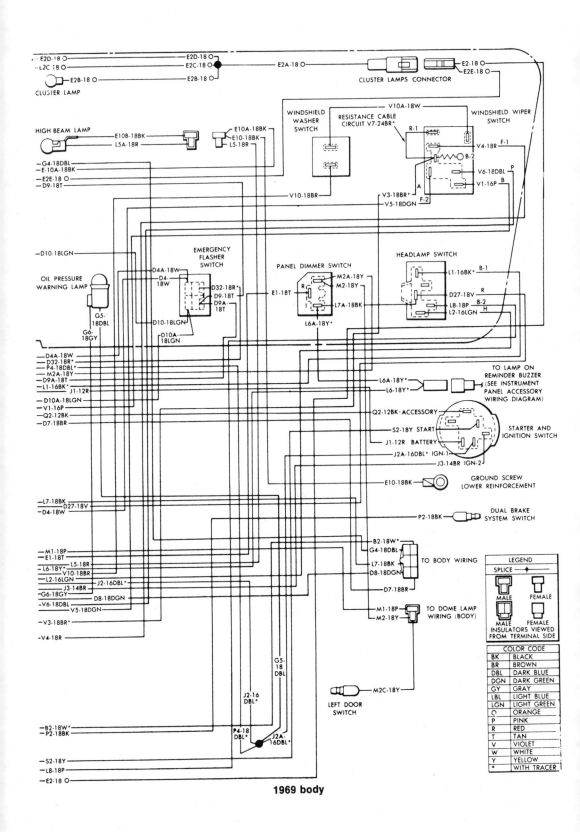

1969 body

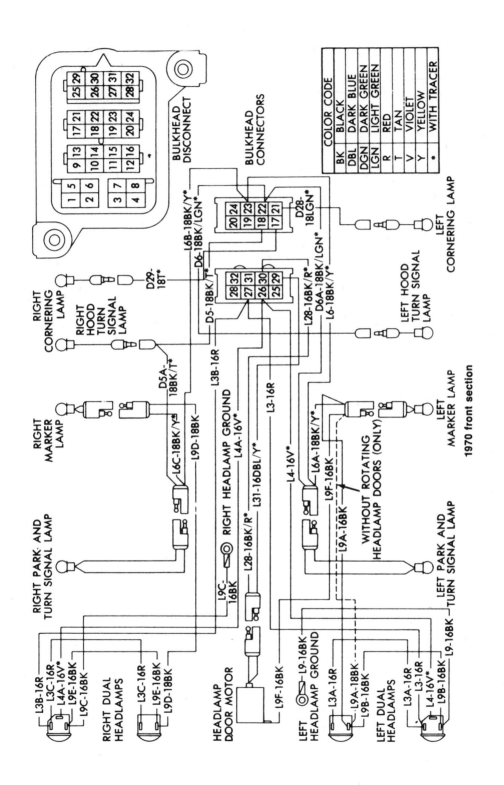

1970 front section

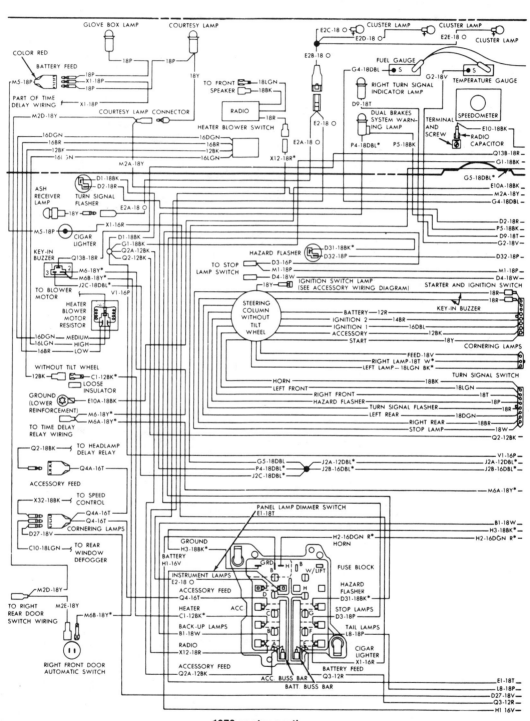

1970 center section

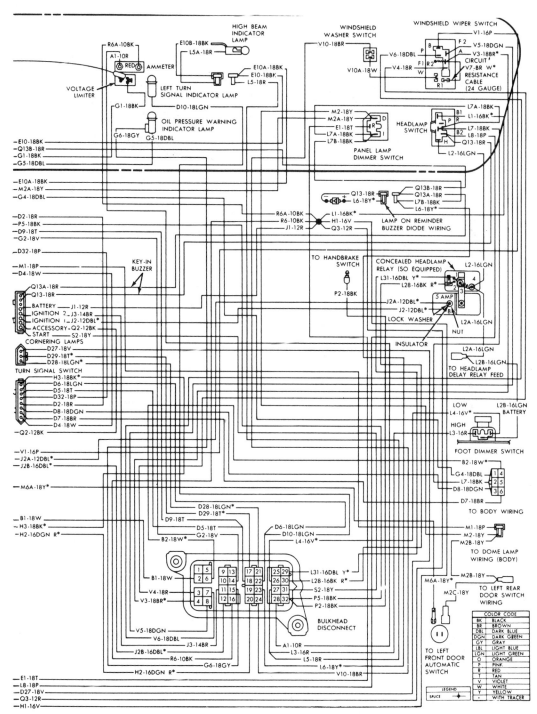

1970 center section

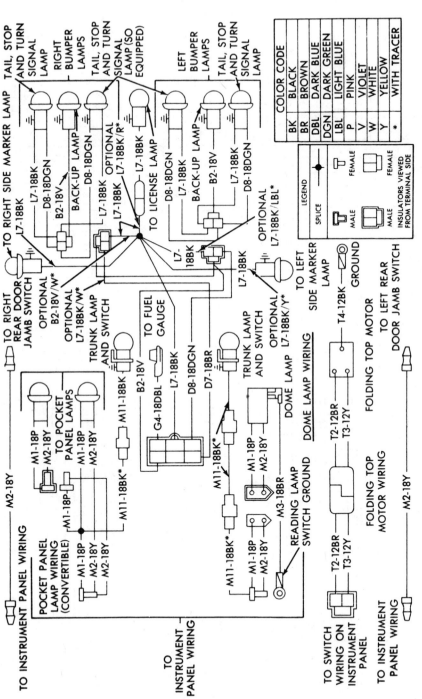

1970 rear section

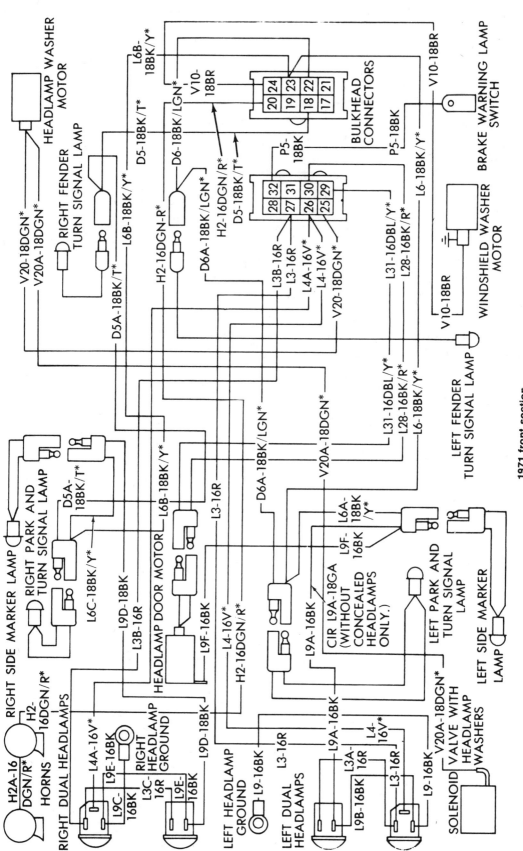

1971 front section

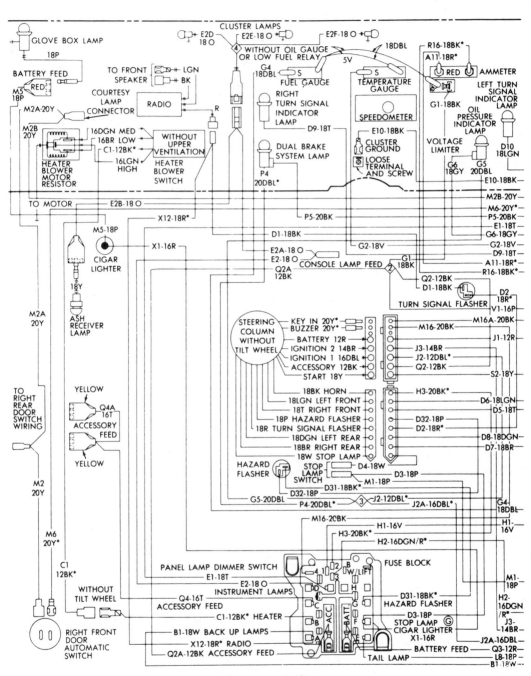

1971 center section

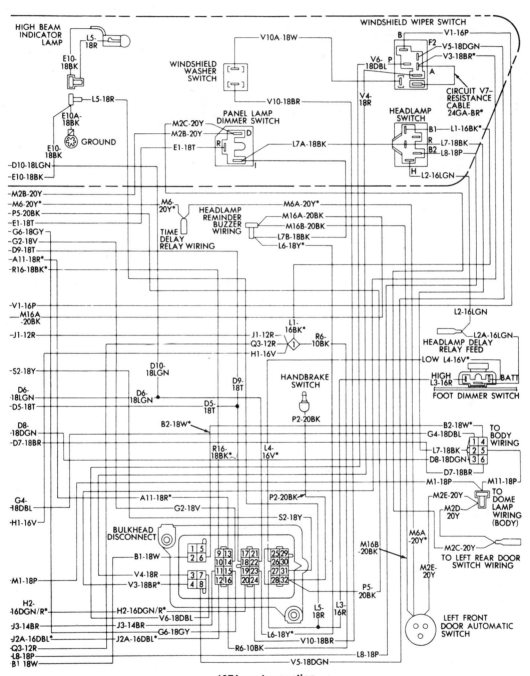

1971 center section

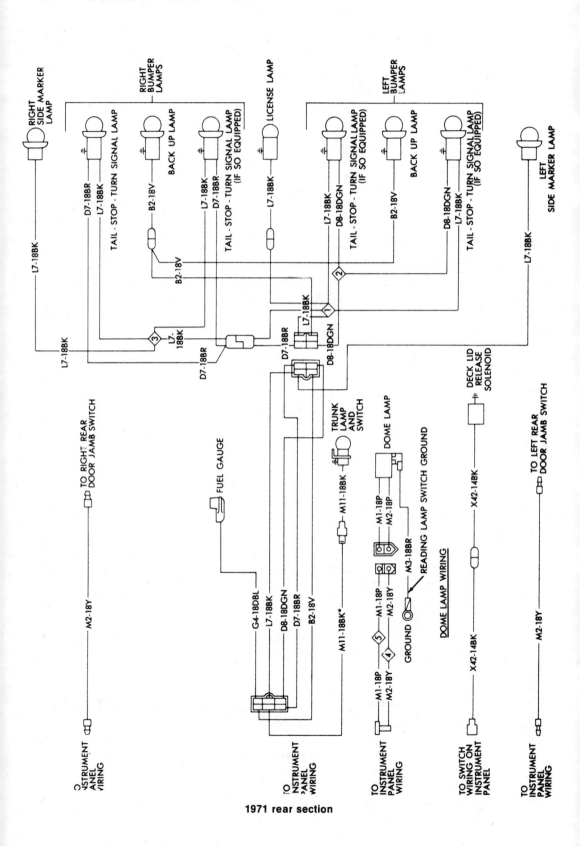

1971 rear section

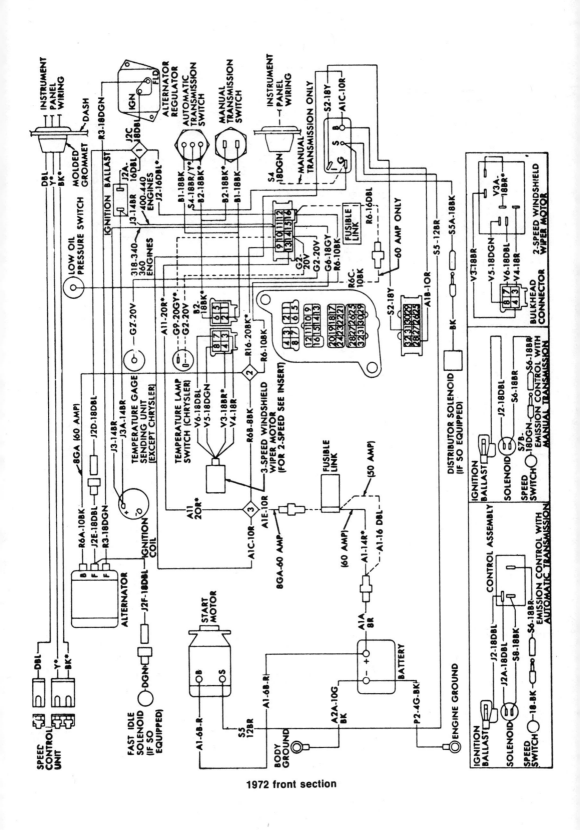

1972 front section

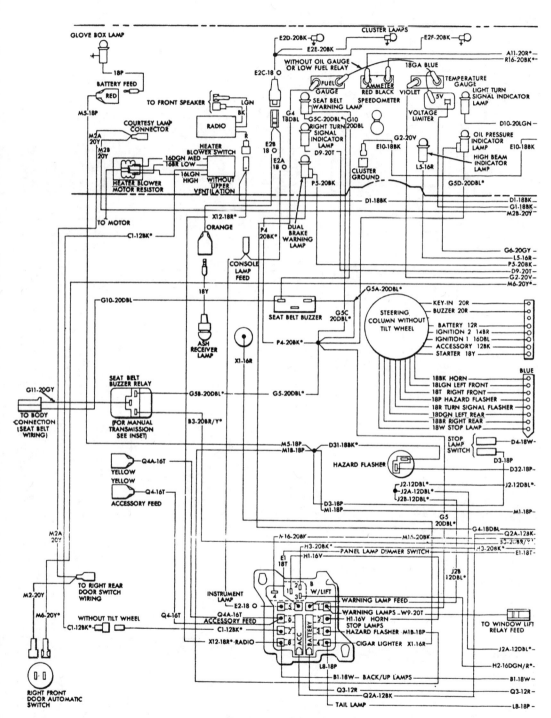

1972–73 center section

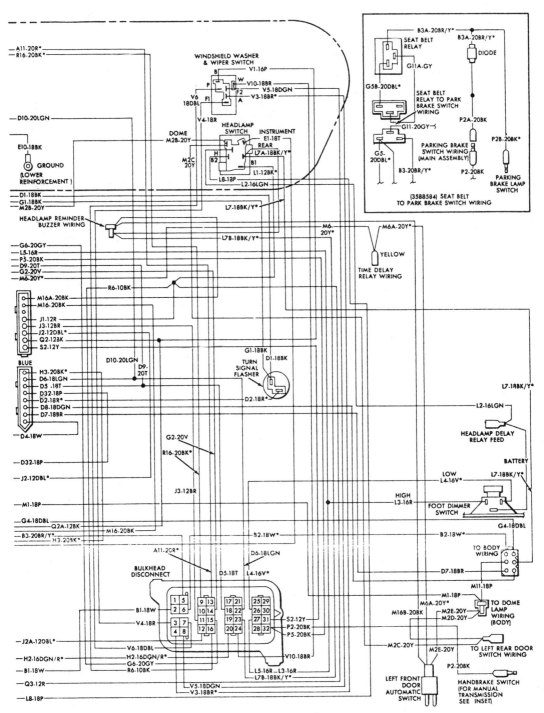

1972–73 center section

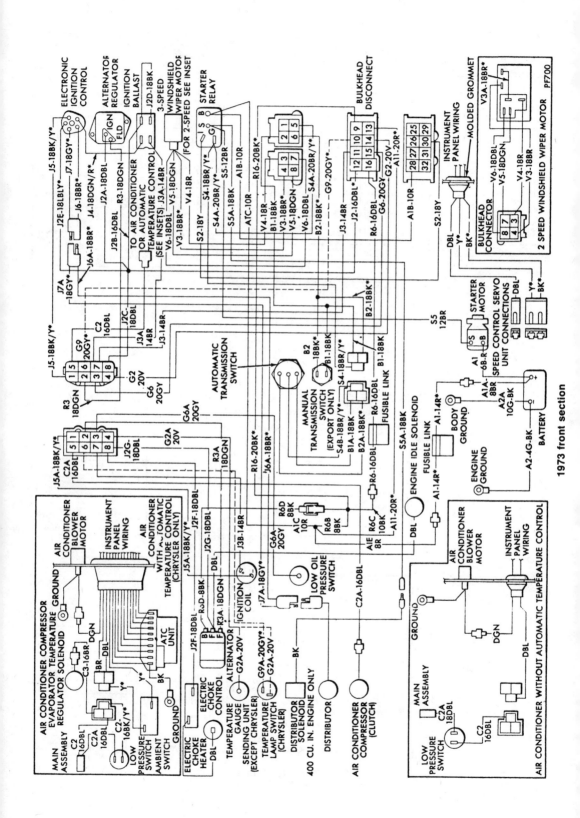

1973 front section

NOTE: *Due to the complexity and length of the 1974–76 wiring diagrams, they are not included in this manual.*

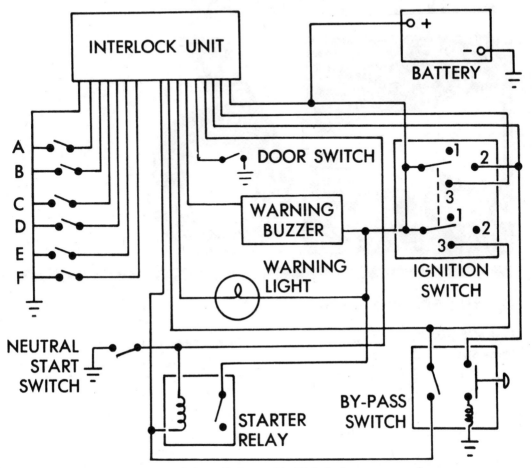

Schematic for seat belt interlock

Clutch and Transmission

MANUAL TRANSMISSION

LINKAGE ADJUSTMENTS

Three-Speed 1968–69

1. Remove the first-reverse rod swivel from steering column and second-third rod swivel from the torque shaft lever.
2. Make sure that transmission shift levers are in neutral (middle detent) position.
3. Adjust the second-third rod swivel by loosening clamp bolt and sliding swivel along rod so it will enter torque shaft lever while hand lever on steering column is held 12 degrees above horizontal position. Install washers and clip. Tighten the swivel clamp bolt to 100 in. lbs.
4. Place a screwdriver or other suitable tool between cross-over blade and second-third lever at the steering column so that both lever pins are engaged by cross-over blade.
5. Adjust the first-reverse rod swivel by loosening locknut and turning swivel so it will enter the first-reverse lever at steering column. Install all washers and the clip. Tighten the swivel lock nut to 70 in. lbs.
6. Remove the tool from cross-over blade at steering column and shift through all gears to check adjustment and cross-over smoothness.

Three-Speed 1970–71

1. Remove both shift rod swivels from the transmission shift levers. Be sure that the transmission shift levers are in neutral (middle) position.
2. Move the shift lever to line up locating slots in the bottom of steering column shift housing and bearing housing. Place a suitable tool in the slot and lock the ignition switch.
3. Place a screwdriver or suitable tool between crossover blade and second-third lever at steering column so that both lever pins are engaged by crossover blade.
4. Set the first-reverse lever on transmission to reverse position (rotate clockwise).
5. Adjust the first-reverse rod swivel by loosening the clamp bolt and sliding swivel along rod. It should enter first-reverse lever at transmission. Install washers and clip. Tighten the swivel bolt to 100 in. lbs.
6. Remove the gearshift housing locating tool, unlock the ignition switch, and shift column lever into Neutral position.
7. Adjust the second-third rod swivel by loosening clamp bolt and sliding

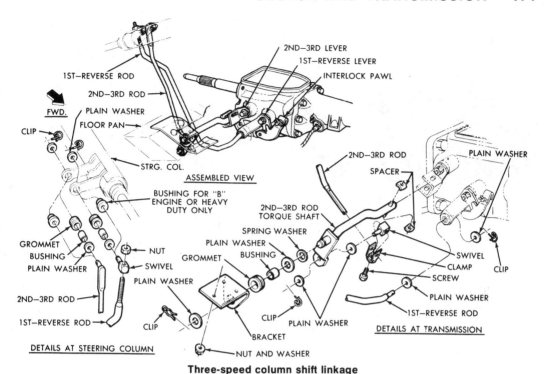

Three-speed column shift linkage

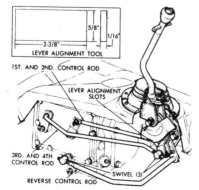

Four-speed floor shift linkage

swivel along rod so it will enter second-third lever at the transmission. Install washers and clip. Tighten the swivel bolt to 100 in. lbs.

8. Remove the tool from crossover blade at steering column, and shift through all gears to check adjustment and cross-over smoothness.

Four-Speed

1. Make up a lever aligning tool from $1/16$ in. thick metal as in illustration.

2. With the transmission in Neutral, disconnect all control rods from the transmission levers.

3. Insert the lever aligning tool through the slots in the levers and against the back plate. This locks the levers in neutral.

4. With all transmission levers in neutral, adjust the length of the control rods so they enter the transmission levers freely without rearward or forward movement.

5. Install the control rod flat washers and retainers. Remove the aligning tool.

6. Check the linkage for ease of shifting into all gears and for ease of crossover.

REMOVAL

All Top Cover Three-Speed Transmissions

1. Drain the transmission.

2. Disconnect the driveshaft at the rear universal joint. Carefully pull the shaft yoke out of the transmission.

3. Disconnect the speedometer cable and the back-up light switch.

4. Install the engine support fixture or jack up the engine about 1 in. and block it in place.

5. Disconnect the transmission extension housing from the center crossmember.

6. Support the transmission with a jack and remove the crossmember. Re-

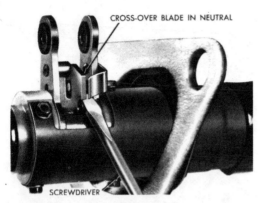

CROSS-OVER BLADE IN NEUTRAL

SCREWDRIVER

Hold the crossover blades in neutral with a screwdriver

move the bolts that attach the transmission to the clutch housing.

7. Slide the transmission rearward until the pinion shaft clears the clutch disc before lowering the transmission.

8. Lower the transmission and remove it.

Fully Synchromesh, Side Cover Three-Speed

1. Remove the shift rods from the transmission levers.

2. Drain the transmission fluid.

3. Disconnect the drive shaft at the rear universal joint. Mark both parts for reassembly.

4. Carefully pull the yoke out of the transmission extension.

5. Disconnect the speedometer and backup lights.

6. Remove part of exhaust system if it blocks transmission.

7. Raise the engine slightly and block it in place.

8. Support the transmission with a jack, and remove crossmember.

9. Remove the transmission to clutch housing bolts.

10. Slide the transmission to rear until the drive pinion shaft is clear. Clear the clutch disc, lower transmission, and remove it from vehicle.

Four-Speed

1. Raise the vehicle on a hoist and drain the transmission.

2. Disconnect all shift controls from the transmission levers. Remove all

three bolts securing the shift unit to extension housing.

3. Disconnect the driveshaft at the rear universal joint. Carefully pull the yoke out of the transmission extension.

4. Disconnect the speedometer cable and backup light switch leads.

5. Disconnect the left exhaust pipe or dual exhausts. Disconnect the parking brake cable.

6. Raise the engine slightly and block it in place.

7. Disconnect the transmission extension from the crossmember.

8. Remove the crossmember.

9. Support the transmission with a jack. Remove the clutch housing to transmission bolts.

10. Slide the transmission rearward until the drive pinion shaft clears clutch disc.

11. Lower the transmission and remove it from the vehicle.

INSTALLATION—ALL

Lightly grease the inner end of the pilot shaft bushing in the flywheel. In addition, grease the pinion bearing retainer pilot at the clutch release shaft.

Position the transmission so that the drive pinion is centered in the clutch housing bore. Push the transmission forward until the pinion shaft enters the clutch disc. Place the transmission in gear. Twist the output shaft until the splines align. Push the transmission forward until it is seated against the clutch housing.

CAUTION: *The transmission must not hang after the pinion is inside the clutch.*

Replace the transmission coupling bolts. Torque them to 50 ft lbs. With the aid of a drift, align the crossmember bolt-holes and install and torque the bolts to 40–50 ft lbs. Remove the engine support fixture and hooks. Install the extension housing and bolt in position. If so equipped, tighten the engine mount-to-cross-member bolt. Install and perform the gearshift linkage adjustment. Connect the driveshaft and universal joints. Connect the exhaust system and fill the transmission with the appropriate lubricant. Road-test the vehicle.

CLUTCH

All models utilize a single, dry-plate type of clutch which is operated by a pedal suspended under the dash. All models are equipped with a return spring; some models have centrifugal rollers assembled betweeen the pressure plate and cover. Six-cylinder and light-duty V8 models utilize a non-centrifugal type of clutch; six-cylinder heavy-duty usage and most V8s use a semi-centrifugal type.

CLUTCH LINKAGE (HEIGHT AND FREE-PLAY) ADJUSTMENT

1. If the vehicle is equipped with a gearshift interlock rod (six-cylinder models and some light-duty V8s with three-speeds), disconnect it by loosening the rod swivel clamp screw.

2. Adjust the fork rod by rotating the self-locking nut to provide $5/32$ in. free-play at the end of the fork. This adjustment will result in the proper one-inch free-play at the clutch pedal.

3. If the gearshift interlock was disconnected, refer to its adjustment procedure below.

GEARSHIFT INTERLOCK ADJUSTMENT

1. Disconnect the interlock pawl from the clutch rod swivel.

2. Adjust clutch pedal free-play.

3. With the first-reverse lever of the transmission in the neutral (middle detent) position, the interlock pawl should enter the slot in the first-reverse lever.

4. Loosen the swivel clamp bolt and move the swivel on the rod to enter the pawl. Install the washers with a clip. Hold the interlock pawl forward and torque the swivel clamp bolt to 100–125 in. lbs. The clutch pedal must be in the fully returned position during this adjustment. Under no circumstance should the clutch rod be pulled rearward to engage the pawl swivel.

5. Shift the clutch through all of the gear positions at least three times, clutch action should be normal.

6. Disengage the clutch and shift half-way to first or reverse gear. The clutch should be held down by the interlock within 1–2 in. of floor.

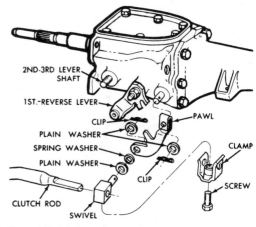

Gearshift interlock assembly

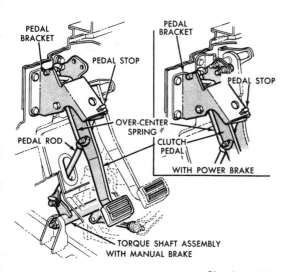

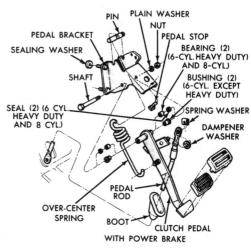

Clutch pedal and linkage

REMOVAL AND INSTALLATION

1968–71

1. Remove the transmission.

2. Remove the clutch housing pan.

3. Disconnect the fork return spring from the clutch housing and release the fork.

4. Take off the spring washer fastening fork rod-to-torque shaft lever pin. Remove the pin from the rod and release the fork.

5. On those models with three-speed transmissions (if this procedure is applicable to the vehicle in question) remove the clip and the plain washer which secures the interlock rod-to-torque shaft lever and remove the washers and rod from the torque shaft.

6. Remove the sleeve assembly and clutch release bearing from the clutch release fork.

7. Punch-mark the clutch cover and flywheel so they may be installed in their same relative positions.

8. Loosen the clutch cover attaching screws in two stages to avoid bending the cover flange.

9. Remove the clutch assembly. Be careful not to contaminate the clutch with grease or oil.

Clutch installation is performed in the following order:

1. Lightly lubricate the drive pinion bushing in the end of the crankshaft with ½ teaspoon of long-life chassis grease. The lubricant should be inserted in the radius in back of the bushing.

2. Thoroughly clean the surfaces of the flywheel and pressure plate with fine sandpaper. All oil or grease must be removed at this time.

3. Position the clutch disc, pressure plate, and cover in the mounting position. The springs on the disc damper must be facing away from the flywheel. Do not touch the disc facing at any time. Insert a clutch disc aligning arbor or suitable substitute (such as a spare transmission drive pinion) through the disc hub and into the bushing.

4. Align the punch marks that were

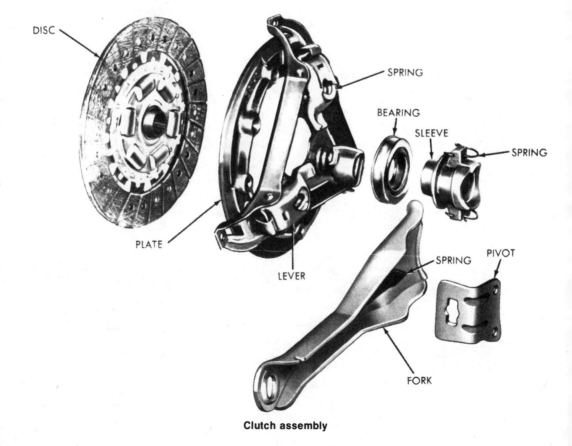

Clutch assembly

Make punch marks to align the clutch cover and flywheel

made at removal. Install the clutch cover bolt but do not tighten it.

5. Tighten all bolts a few turns at a time in an alternate sequence. Torque ⁵/₁₆ in. bolts to 200 in. lbs and ⅜ in. bolts to 30 ft lbs. Remove the alignment tool.

6. Pack the bearing sleeve cavity with an appropriate NLGI Grade 2 EP grease. Apply the same lubricant to the release fork pads of the sleeve.

7. Insert the release bearing and sleeve assembly into the clutch housing as far forward as possible. Lightly lubricate the fork fingers and retaining spring.

8. Insert the fork fingers under the clutch sleeve retaining springs. The retaining springs on the sleeve must have lateral freedom.

9. Make sure that the groove in the seal is properly seated in the seal-opening flange in the clutch housing. Replace the pedal rod on the torque shaft lever pin and secure it with spring washer.

10. Insert the threaded end of the fork rod assembly into the opening provided in the end of the release fork rod. Replace the eye end of the fork rod on the torque shaft lever pin and lock it in place with a spring washer.

11. If applicable, install the fork return spring between the release fork and the clutch housing.

12. If applicable, install the spring and plain washer with the interlock rod in the torque shaft lever and lock it in position with a washer and clip.

13. When installing the transmission,

be careful that no grease settles on the splines or pilot end of the transmission drive pinion.

14. Install the transmission and adjust the clutch pedal free-play.

AUTOMATIC TRANSMISSION

There are two Chrysler TorqueFlite® transmissions. The A-904 is standard on the 225 and 318. The A-727 is the heavy-duty version TorqueFlite and is an option on the 225 and 318 and is standard on all other engines.

PAN REMOVAL AND FLUID CHANGE

1. Raise the car to gain working clearance underneath it. Be sure that it is securely supported with jackstands.

2. Place a large, wide-mouthed container underneath of the transmission oil pan.

3. Loosen, but do not remove, the oil pan bolts. Tap the pan free at one corner and allow the fluid to drain into the container.

4. Remove the converter access plate and remove the converter drain plug so that the fluid in the converter will drain into the container.

Converter drain plug

5. Install the converter drain plug and tighten it to 90 in. lbs. Secure the access plate with its mounting screw.

6. Adjust the Reverse band and install a new filter, if necessary (see below).

7. Clean and install the oil pan, using a new gasket. Tighten the pan bolts to 150 in. lbs.

8. Add 6 qts of DEXRON® automatic transmission fluid through the filler tube.

9. Start the engine and allow it to run at idle for at least 2 min. Apply the parking brake and engage each gear

range for several seconds. Leave the selector in Neutral (N).

10. Add enough fluid to bring the level on the dipstick up to the "ADD" mark. Run the engine until it reaches normal operating temperatures; the fluid level should fall between the "ADD" and "FULL" marks. Add fluid as necessary; do not overfill.

FILTER REPLACEMENT

1. Remove the pan as described above.

2. Remove the three filter attaching bolts and remove the old filter.

3. Install a new filter, making sure all gaskets are in place.

4. Install the pan using a new gasket.

BAND ADJUSTMENTS

Kick-Down Band

The kick-down band adjusting screw is located on the left-hand side of the transmission case near the throttle lever shaft.

Kick-down band adjusting screw

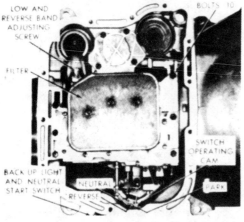

Oil pan removed—showing Low/Reverse band adjustment

1. Loosen the locknut and back it off about five turns. Be sure that the adjusting screw is free in the case.

2. Using a torque wrench and, if necessary, an adapter, torque the adjusting screw to 50 in. lbs (if an adapter is used) or to 72 in. lbs (if an adapter is not used).

3. Back off the adjusting screw the exact number of turns specified below. Keep the screw from turning and torque the locknut.

Low and Reverse Band

The oil pan must be removed from the transmission to gain access to the Low/Reverse band adjusting screw.

1. Drain the transmission and remove the oil pan.

2. Loosen the band adjusting screw locknut and back it off about five turns. Be sure that the adjusting screw turns freely in the lever.

3. Using a torque wrench and, if necessary, an adapter, torque the adjusting screw to 47–50 in. lbs (if an adapter is used) or to 72 in. lbs (if an adapter is not used).

4. Back off the adjusting screw the exact number of turns specified below. Keep the screw from turning and torque the locknut to 30 ft lbs.

5. Install the pan and use a new gasket. Torque the bolts to 150 in. lbs. Refill the transmission with DEXRON® transmission fluid.

Band Adjustment Specifications

	KICK-DOWN BAND	
Year	Engine/Trans	Turns
1968–76	All engines/A-904	2
1968–70	All engines/A-727 transmission	2
1971–72	All engines/A-727 transmission	2½
1973–76	All engines/A-727 transmission	2½

	LOW AND REVERSE BAND	
Year	Engine/Trans	Turns
1968–71	All engines except 318/A-904 transmission	3¼
1968–76	318/A-904 transmission	4
1968–70	All engines/A-727 transmission	4
1971–76	All engines/A-727 transmission	2

Neutral Safety Switch Repair

The neutral safety switch is mounted in the transmission case on all models. When the gearshift lever is placed in either the Park or Neutral position, a cam, which is attached to the transmission throttle lever inside the transmission, contacts the neutral safety switch and provides a ground to complete the starter solenoid circuit.

The back-up lamp switch is incorporated into the neutral safety switch. This combination switch can be identified by the three electrical terminals on the rear of the switch. On this type of switch, the center terminal is for the neutral safety switch and the two outer terminals are for the back-up lamps.

There is no adjustment provided for the neutral safety switch. If a malfunction occurs, first check to make sure that the transmission gearshift linkage is properly adjusted and that the actuator cam is centered in the switch mounting hole in the transmission. If the malfunction continues, the switch must be removed and replaced.

To remove the switch, disconnect the electrical leads and unscrew the switch from the transmission. Use a drain pan to catch the transmission fluid that drains out of the mounting hole. Install a new switch using a new gasket and refill the transmission to the proper level.

SHIFT LINKAGE ADJUSTMENT

1968-69 Console Shift and Column Shift

1. Place the gearshift lever in the Park position.
2. Loosen the lower rod swivel clamp screw several turns.
3. Move the transmission control lever to the rearmost, or Park, detent.
4. With the gearshift lever in the Park position and the transmission lever in the Park position, tighten the swivel clamp screw and test the linkage.
5. The shift effort must be free and the detents should feel crisp. All gate stops must be positive. It should be possible to start the car in the Park and Neutral positions only.

1970-76 Console Shift

1. Loosen the clamping screw in the adjustable rod swivel clamp.
2. Line up the locating slots in the bottom of the shift housing and the bearing housing, found at the upper end of the steering column.

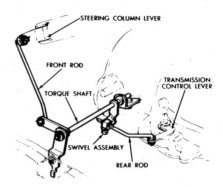

Column shift linkage—1968-69

3. Install the tool to keep the slots aligned and lock the column with the ignition key.
4. Put the console gearshift lever in Park.
5. Put the transmission lever in the rearmost, or Park, position.
6. Set the adjustable rods to the proper length with no load applied in either direction on the linkage and tighten the clamp bolts to 125 in. lbs.

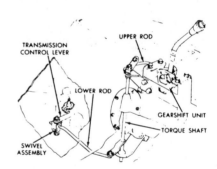

Console shift linkage—all years

7. The shift effort must be free and the detents should feel crisp. All gate stops must be positive. It should be possible to start the car in the Park and Neutral positions only.

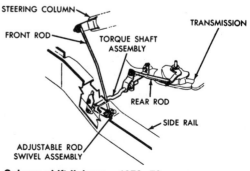

Column shift linkage—1970-76

1970–76 Column Shift

1. Loosen the clamping screw in the adjustable rod swivel clamp.

2. Put the gearshift selector lever in the Park position and lock the steering column with the key.

3. Put the transmission shift control lever in the rearmost, or Park, detent.

4. Set the adjustable rod to the proper length and install the rod with no load in either direction on the linkage. Tighten the clamp bolts 125 in. lbs.

5. Check the linkage as in Step 7 above.

THROTTLE LINKAGE ADJUSTMENT

See Chapter 4, "Emission Controls and Fuel System," for the throttle linkage adjustment procedures.

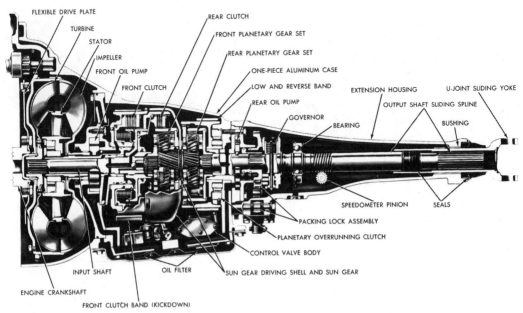

Cutaway view of the TorqueFlite® automatic transmission

Drive Train

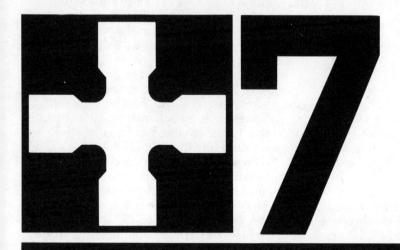

DRIVELINE

Driveshaft and U-Joints
REMOVAL AND INSTALLATION

1. Matchmark the driveshaft, U-joint and pinion flange before disassembly. These marks must be realigned during reassembly to maintain the balance of the driveline. Failure to align them may result in excessive vibration.

2. Remove both of the clamps from the differential pinion yoke and slide the driveshaft forward slightly to disengage the U-joint from the pinion yoke. Tape the two loose U-joint bearings together to prevent them from falling off.

CAUTION: *Do not disturb the bearing assembly retaining strap. Never allow the driveshaft to hang from either of the U-joints. Always support the unattached end of the shaft to prevent damage to the joints.*

3. Lower the rear end of the drive-

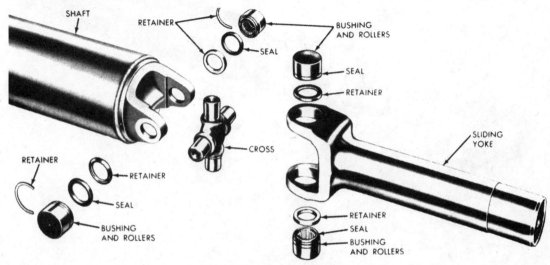

Front cross and roller U-joint

shaft and gently slide the front yoke/ driveshaft assembly rearward disengaging the assembly from the transmission output shaft. Be careful not to damage the splines or the surface which the output shaft seal rides on.

4. Check the transmission output shaft seal for signs of leakage.

5. Installation is the reverse of removal. Be sure to align the matchmarks.

U-Joints

OVERHAUL

1. Remove the driveshaft.

2. To remove the bearings from the yoke, first remove the bearing retainer snap rings located at the base or open end of each bearing cap.

3. Pressing on one of the bearings, drive the bearing in toward the center of the joint. This will force the cross to push the opposite bearing out of the universal joint. This step may be performed using a hammer and suitable drift or a vise and sockets or pieces of pipe. However installation of bearing must be done using the vise or a press.

4. After the bearing has been pushed all the way out of the yoke, pull up the cross slightly and pack some washers under it. Then press on the end of the cross from which the bearing was just removed to force the first bearing out of

the yoke. Repeat Steps 3 and 4 to remove the remaining two bearings.

5. If a grease fitting is supplied with the new U-joint assembly, install it. If no fitting is supplied, make sure that the joint is amply greased. Pack grease in the recesses in the end of the cross.

6. To reassemble start both bearing cups into the yoke at the same time and hold the cross carefully in the fingers in its installed position. Be careful not to knock any rollers out of position.

7. Squeeze both bearings in a vise or press, moving the bearings into place. Continually check for free movement of the cross in the bearings as they are pressed into the yoke. If there is a sudden increase in the force needed to press the bearings into place, or the cross starts to bind, the bearings are cocked in the yoke. They must be removed and restarted in the yoke. Failure to do so will greatly reduce the life of the bearing. Repeat Steps 6 and 7 to reinstall the remaining two bearings.

REAR AXLE

There are two types of Chrysler rear axles, the C type and the non-C type. The axle shafts on the C type are retained by C-shaped locks, which fit in grooves at the inner end of the axle shaft. Axle shafts of the non-C type are retained by a plate

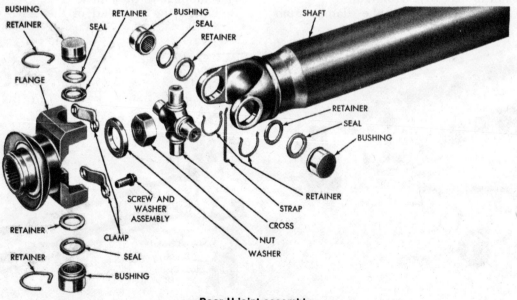

Rear U-joint assembly

located at the brake drum end of the axle or by the backing plate itself. However, visual inspection is necessary to determine the type as the 8¼ in. and 1974–76 9¼ in. axles are the C type.

Axle Shaft

REMOVAL AND INSTALLATION

9¼ (1974–76) and 8¼ Axles—C Type

To visually determine the axle type, remove the differential cover.

1. Raise the vehicle and remove the wheels. Drain the oil from the rear axle.

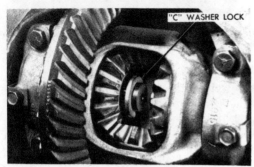

"C" WASHER LOCK

9¼ and 8¼ in. axles with C-locks

2. Remove the differential cover, if not already removed.

3. Remove the differential pinion shaft lockscrew and the differential pinion shaft.

Pinion shaft lockscrew removal and installation

4. Push the flanged end of the axle shaft toward the center of the vehicle and remove the C-lock from the end of the shaft.

5. Remove the axle shaft from the housing being careful not to damage the oil seal. The axle shaft may not slide easily out of the housing. If so, obtain an axle puller.

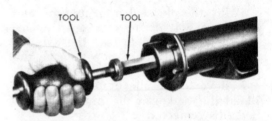

TOOL   TOOL

Removing the axle shaft with a puller—8¼ and 9¼ in. axles

6. The axle bearing will come out with the axle and will have to be replaced at an automotive repair shop.

7. Pry the axle seal loose from the bore and tap a new seal into place.

8. Installation is the reverse of removal. Use new gaskets for reassembly.

NOTE: *1972–76 models with an 8¼ in., or 9¼ in. axle no longer use a paper gasket under the rear axle cover. Instead of the paper gasket, a bead of sealant is now used in production. The sealant is available for service under Part No. 3683829. The sealer should be applied as follows:*

1. Scrape away any remains of the paper gasket.

2. Clean the cover surface with mineral spirits. Any axle lubricant on the cover or axle housing will prevent the sealant from taking.

3. Apply a ¹⁄₁₆–³⁄₃₂ in. bead of sealant to the clean, dry, cover flange. Apply the bead in a continuous bead along the bolt circle of the cover, looping inside the bolt holes as shown.

4. Allow the sealant to air dry.

5. Clean the carrier gasket flange and air dry. Install the cover. If, for any reason, the cover is not installed within 20 minutes of applying the sealant, remove the sealant and start over.

8¼-Non-C Type

1. Jack the car up so that the rear wheels are off the ground. Remove the wheel and brake drum from the axle being removed.

2. Remove the nuts holding the retainer plate to the backing plate. Disconnect the brake line, if necessary.

3. Remove the retainer and reinstall the nuts fingertight to prevent the backing plate from being dislodged.

4. Using a slide hammer, pull out the axle shaft and bearing assembly.

5. Pry the old seal from its bore and install a new seal.

6. You will have to take the axle assembly to a machine shop to have the bearing removed.

7. Reverse the above steps to install. Bleed the brakes if the brake line has been disconnected.

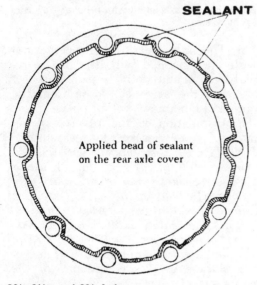

SEALANT

Applied bead of sealant on the rear axle cover

8¾, 9¼, and 9¾ Axles

These axles are all of the non-C type (except for the 1974–76 9¼ in., see as above). They also have both inner and outer axle seals and adjustable end-play.

1. Remove the wheel and brake drum.

2. If the backing plate must be removed, disconnect the brake line at the wheel cylinder.

3. Working through the hole in the flange, remove the five nuts from the retainer plate. The right-hand shaft, with the end-play adjuster in the retainer plate, will also have a lock on one of the studs that will have to be removed at this time.

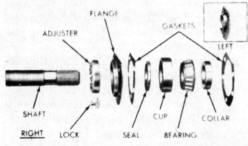

FLANGE
ADJUSTER
GASKETS
LEFT
SHAFT
RIGHT   LOCK   SEAL   CUP   BEARING   COLLAR

Axle shaft components—8¾ in. axle (9¼ and 9¾ similar)

4. Remove the axle and bearing assembly with an axle puller.

5. Pry the inner axle seal from its bore and tap a new seal in place.

Removing the inner axle seals—8¾, 9¼, and 9¾ in. axles

6. If the outer axle seal is located in the axle retainer it will have to be replaced along with the axle bearing at a machine shop. If the seal is located in the backing plate, tap the seal from the plate and install a new one.

7. Install the left-hand axle. Install the retaining plate using a new gasket.

8. Install the right-hand backing plate using a new gasket.

9. Back off the adjuster on the right axle shaft assembly until the inner face of the adjuster is flush with the inner face of the retainer.

10. Slide the right axle assembly into the housing and install the retaining plate.

11. With a dial indicator mounted on the brake support, tighten the adjuster until there is zero end-play. Then, back off approximately four notches to get 0.013–0.023 in. end-play.

12. Install the adjuster lock and recheck the end-play adjustment. Make sure that both axles are fully seated.

13. Reinstall the brake drums and wheels.

REAR AXLE HOUSING

REMOVAL AND INSTALLATION

1. Raise the car and support the car in front of the springs.

2. Remove the rear wheels.

3. Disconnect and plug the hydraulic brake hose at the left side of the axle.

4. Disconnect the parking brake cable.

5. Disconnect the driveshaft at the differential yoke (see above).

6. Remove the shock absorber lower mounts from the axle.

7. Remove the U-bolts and plates which secure the axle to the springs.

8. Remove the axle from the vehicle.

9. Installation is the reverse of removal. Bleed the brakes after installing the axle.

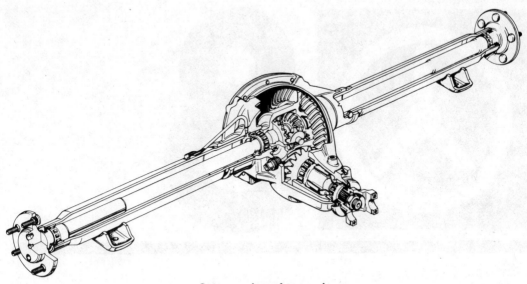

Cutaway view of rear axle

Suspension and Steering

FRONT SUSPENSION

The Plymouth uses torsion bars in place of conventional springs in its front suspension. The rest of the front suspension consists of upper and lower control arms and ball joints. The suspension support members (K-members) and the engine mounts are parts of a sub-frame which is bolted to the front of the unitized body. The pivot shafts and struts of the lower control arms are attached to the legs of the K-member.

The rear anchor points of the torsion bars are part of the engine's rear support, while the front anchors are part of the lower control arms. The front anchors are designed to permit front suspension height adjustment. Caster and camber adjustments are provided for by cams located on the upper control arm pivot bolts.

The ball joints are of the semi-permanent, lubrication type, and are integral with the steering arms. See the first chapter of the book for the proper lubrication intervals.

All torsion bars and ball joints use balloon-type seals at the front and rear anchor points to keep out water and foreign material. These seals, along with the tie rod seals, should be inspected for damage at every oil change.

All suspension points which are rubber-mounted should be tightened only when the suspension is at the specified height and with the full weight of the car on its wheels.

CAUTION: *Never lubricate the rubber bushings.*

The 1974–76 Plymouth front suspension has been revised to include new lower control arms which are more serviceable. The new lower control arm ball joints are of the "press-fit" tension type. A new alignment adjustment system and more serviceable struts have also been incorporated.

Torsion Bars

REMOVAL AND INSTALLATION

CAUTION: *Contrary to appearance, the torsion bars are not interchangeable from the right to the left side. They are marked with an "R" or an "L" stamped at one end to aid in their correct positioning.*

1. Take the rebound bumper off the mounting bracket under the upper control arm.

2. Raise the car by the *body* only, so

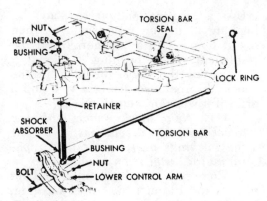

Torsion bar

that the front suspension has no load on it. If jacks are used to raise the car, place a support under the center K-member first, then jack the car up.

CAUTION: *Never raise the car by placing a jack under an unsupported K-member. Be sure that the car is securely supported.*

3. Back-off (counterclockwise) the anchor adjusting bolt to release the load on the torsion bar.

4. Withdraw the seal from the rear anchor and remove the anchor lockring.

5. Install a clamp toward the rear of the torsion bar and move the torsion bar backward by striking the clamp with a hammer.

CAUTION: *Do not apply heat to the torsion bar or to either of the anchors. Use care not to mar the surface of the torsion bar during removal.*

6. Remove the clamp and carefully slide the rear anchor seal off the front end of the torsion bar.

7. Slide the torsion bar out through the rear anchor.

Install the torsion bar in the following sequence.

1. Clean the hex openings of both the front and rear anchors as well as the male ends of the torsion bar.

2. Check the balloon seal for signs of damage and replace it if required.

3. Check the torsion bar for any signs of scores or nicks. If any are present, dress them down, in order to remove all sharp edges. Paint the repaired areas with rust preventive paint.

4. Replace the torsion bar adjusting nut if it is corroded. Lubricate it for smooth operation.

5. Fit the torsion bar through the rear anchor. Slide the rear anchor seal over the torsion bar with its large cupped end toward the rear of the bar.

6. Use multi-purpose grease to coat both of the hex ends of the bar.

7. Slip the torsion bar into the lower control arm hex opening. Seat the lockring in its groove on the rear of the anchor.

8. Fill the annular opening of the rear anchor with multipurpose grease. Fit the lip of the seal into its groove in the rear anchor.

9. Rotate the adjustment bolt clockwise to load the torsion bar.

10. Lower the car and perform the front suspension height adjustment (see below).

11. Fit the upper control arm rebound bumper and tighten its securing nut to 200 in. lbs.

Shock Absorbers
REMOVAL AND INSTALLATION

1. Remove the nut and the retainer from the top end of the shock absorber piston rod. (This must be done from inside the engine compartment on some models.)

2. Lift up the car so that its wheels do not touch the floor. *Be sure that it is securely supported.*

3. Unfasten the nut on the lower attachment bolt, and withdraw the bolt completely (1968–73).

On 1974–76 models, remove the nut, retainer, and bushing from the lower control arm.

4. To compress the shock absorber

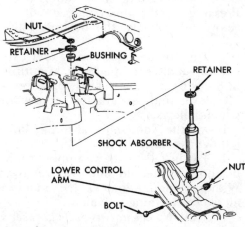

Front shock—1968–73

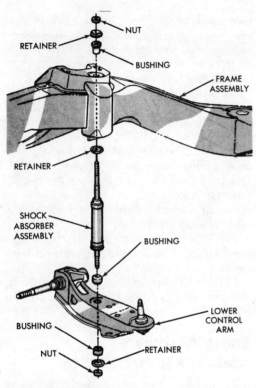

NUT

RETAINER

BUSHING

FRAME ASSEMBLY

RETAINER

SHOCK ABSORBER ASSEMBLY

BUSHING

BUSHING

NUT

RETAINER

LOWER CONTROL ARM

Front shock—1974–76

fully, push up on it. Remove it from the car, while compressed, by pulling it downward, and out of its upper mounting bushing.

5. Inspect the upper mounting bushing. If it shows signs of wear or deterioration, remove it by pressing it out of the inner sleeve. Then pry or cut out the rubber bushing.

NOTE: *Once the bushing is removed, it must be replaced with a new one.*

6. Remove the lower bushing, if necessary, by pressing on its *outer* sleeve. Pressing on the inner sleeve will not completely remove the bushing (1968–73).

7. Test the shock absorber and expel any air which is trapped in it, as outlined in Steps 6–10 of the rear shock absorber "Removal and Installation" procedure which follows.

Install the front shock absorber in the following manner.

1. To install a new upper rubber bushing, if the old one was removed. Withdraw the inner steel sleeve and immerse the bushing in water.

CAUTION: *Do not use oil, grease, or soap to lubricate the bushing.*

Use a twisting motion to start the bushing in the mounting bracket, then *tap* it home with a hammer. Insert the steel sleeve in the bushing.

2. Install the lower bushing in the shock absorber eye, using a vise and a drift, if it was removed (1968–73).

NOTE: *New shocks come with the lower bushing already installed, but the bushings may be obtained separately for installation in old shocks.*

3. Compress the shock fully, and insert its rod through the upper bushing. Fit the upper retainer and tighten the upper attachment nut to 25 ft lbs.

4. On 1968–73 models, align the eye of the shock absorber with the lower control arm mounting holes. Install the bolt and tighten the retaining nut to 50 ft lbs.

5. On 1974–76 models, slip the upper bushing over the eye of the lower bayonet, push up on the shock to clear the control arm, and fit the bayonet through the hole in the lower control arm. Install the lower bushing, and retainer over the bayonet. Tighten the lower retaining nut to 40 ft lbs with the full weight of the car on the wheels.

Lower Ball Joint

INSPECTION

1. Lift up the front of the car. Install floor stands, as a safety precaution, under both of the lower control arms. The stands should be mounted as far outboard as possible to prevent the upper control arms from contacting their rubber rebound bumpers.

2. With the car's weight on the control arms, install a dial indicator and clamp assembly on the lower control arm. Position the plunger end of the dial indicator

Measuring lower ball joint axial travel

against the ball joint housing. Zero the dial indicator.

3. Measure the axial travel of the ball joint housing arm, in relation to the stud, by raising and lowering the wheel.

NOTE: *A pry bar, placed under the center of the tire, is useful for this operation.*

4. If the axial play exceeds 0.020 in., on 1974–76 models, or 0.070 in., on 1968–73 models, the lower ball joint requires replacement.

REMOVAL AND INSTALLATION

1968–73

The lower compression-type ball joint and the steering arm, used on these models, are an integral unit. Because of this, they cannot be replaced separately.

1. Take the upper control arm rebound bumper off.

2. Raise the car. Be sure that the suspension is under no load (full rebound).

CAUTION: *If jacks are used, there must be a support placed between the K-member and the jack.*

3. Back-off (counterclockwise) the torsion bar adjuster to remove the load on the torsion bar.

4. Remove the wheel, tire, and the drum as an assembly. The brake shoe adjuster may have to be loosened, in order to remove the drum.

NOTE: *If the car is equipped with disc brakes, see Chapter 9 for the correct removal procedure.*

5. Unfasten the two lower bolts from the brake support which secure the ball joint/steering arm assembly to the steering knuckle.

6. Remove the end of the tie-rod from the steering arm with a suitable puller.

7. Use a ball joint puller to withdraw the ball joint stud from the lower control arm. The ball joint/steering arm assembly may now be withdrawn completely from under the car.

Perform the installation of the lower ball joint in the following order:

1. Position a new seal (if required) over the ball joint, being certain that the lip of the seal is fully seated in the housing.

2. Attach the ball joint/steering arm assembly to the steering knuckle and tighten the attachment bolts to 120 ft lbs.

3. Fit the ball joint stud into the opening in the lower control arm. Tighten the stud retaining nut to 115 ft lbs. Install the cotter pin. Lubricate the ball joint (See Chapter 1).

4. Check the tie-rod seal for signs of damage and replace it if necessary. Attach the tie-rod end to the steering knuckle arm. Torque its securing nut to 40 ft lbs. Install the cotter pin.

5. Load the torsion bar by rotating its adjusting nut clockwise.

6. Install the wheel and brake assembly. Adjust the front wheel bearing.

NOTE: *Detailed procedures for the above step may be found in Chapter 9.*

7. Lower the car. Fit the upper control arm rebound bumper and tighten its securing nut to 200 in. lbs.

8. Measure the front wheel height and the wheel alignment. Adjust them as outlined below.

1974–76

The ball joints on these models are press-fitted, and may be removed and installed without replacing the steering arm.

1. Using the ignition key, place the steering column lock in the "OFF" or "UNLOCKED" position.

2. Perform Steps 1 through 4 of the "1968–73 Lower Ball Joint Removal" procedure.

3. Remove the cotter pins and nuts which secure the upper and lower ball joint studs.

4. Use a ball joint puller to remove the upper stud from the steering knuckle. Once the puller is installed on the upper stud, strike the steering knuckle with a hammer to loosen the stud. Do not use the puller alone to remove the stud from the steering knuckle.

5. Remove the ball joint from the lower control arm with a puller.

Installation of a new ball joint is performed in the following order:

1. Press-fit the new ball joint into the lower control arm.

2. If required, press a new rubber seal over the ball joint until it is securely locked into position.

3. Fit the ball joint stud into the steering knuckle arm opening and secure it with its retaining nut.

4. Tighten both upper and lower ball joint stud retaining nuts to 135 ft lbs and insert their cotter pins.

5. Lubricate the ball joint as detailed in Chapter 1.

6. Load the torsion bar by rotating its retaining nut clockwise.

7. Install the wheel and brake; then adjust the front wheel bearings as detailed in Chapter 9.

8. Lower the car. Fit the upper control arm rebound bumper and tighten its securing nut to 200 in. lbs.

9. Check the suspension height and wheel alignment (see below).

Upper Ball Joint

REMOVAL AND INSTALLATION

Removing ball joint stud

Perform this procedure with the steering column lock in the "OFF" or "UN-LOCKED" position.

1. Raise the car by placing a floor jack under the lower control arm. Support the car securely with jackstands.

2. Remove the wheel and tire. Remove the brake drum, if so equipped. On models with disc brakes, remove the caliper and disc from the steering knuckle. Position the caliper out of the way with the brake line attached.

3. Remove the nut that attaches the upper ball joint to the steering knuckle. Loosen the ball joint stud from the steering knuckle. Ball joint removal tools are available at auto parts stores and large mail order houses. These press the ball joint out. Follow the manufacturer's directions for operating the tool. An alternate method for stud removal is hammering firmly on the side of the steering knuckle until the stud pops out of the steering knuckle. Never strike the ball joint stud. Brace the knuckle with another hammer.

4. Unscrew the upper ball joint from the upper control arm and remove it from the vehicle.

Install the upper ball joint by using the following procedure.

1. Screw the ball joint into the control arm, as far as possible, by hand.

CAUTION: *If the original control arm is used, be sure that the ball joint threads engage squarely with those on the arm. The balloon-type seals must be replaced with new ones once they are removed.*

2. Continue tightening until the ball joint bottoms in its housing. It must be tightened to a *minimum* of 125 ft lbs on 1968–73 models and to 150 ft lbs on 1974–76 models. If the torque figure cannot be obtained, check the threads on the ball joint and the control arm. Replace either one, or both of them, if they are worn.

3. Position a new seal on the ball joint stud and install it on the ball joint. Be sure that the seal is fully seated in the ball joint housing.

4. Fit the ball joint into the steering knuckle and tighten its securing nuts to 100 ft lbs on 1968–73 models and on 1974–76 models to 135 ft lbs. Install the cotter pin and lubricate the ball joint. If the replacement ball joint has a "knock-off" grease fitting, break off the portion over which the grease gun was installed.

5. Install the lower stud nut and tighten it to 115 ft lbs on 1968–73 models and to 135 ft lbs on 1974–76 models.

6. Install the wheel and brake assemblies. Adjust the front wheel bearing.

NOTE: *Detailed procedures for the above step may be found in Chapter 9.*

7. Lower the car. Adjust the front suspension height as detailed below.

Lower Control Arm Assembly

REMOVAL AND INSTALLATION

1968–73

1. Remove the wheel, tire, and brake assembly, as outlined in Chapter 9. Un-

fasten the lower shock absorber securing bolt and push the shock up, out of the way.

2. Raise the car and follow the procedure outlined in the torsion bar removal from the lower control arm (Steps 1–5).

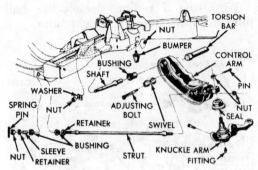

Lower control arm assembly—1968–73

3. Detach the tie-rod end from the steering knuckle arm with a suitable puller. Be careful not to damage its seal.

4. Remove the stabilizer bar strut strap from the lower control arm (if so equipped).

5. Unfasten the bolts which attach the steering knuckle arm to the brake support. Position the brake support out of the way.

6. Withdraw the ball joint stud from the lower control arm with a puller.

CAUTION: *Be careful to avoid damaging the seal.*

7. Remove the strut spring pin (if so equipped), the front nut, and the bushing from the forward end of the K-member.

8. Unfasten the cotter pin, the nut, and the washer from the lower control arm.

9. Using a soft-ended hammer, tap the lower control arm shaft to remove it from the K-member. Withdraw the lower control arm, shaft, and strut as a complete assembly.

10. The strut bushing need only be removed from the K-member if it is damaged. Remove the strut bushing inner retainer from the strut.

11. With the strut portion of the control arm assembly in a vise, remove its nut, and then the strut itself. Unfasten the torsion bar adjusting bolt and swivel. Remove the jounce bumper (if so equipped).

12. Position the lower control arm as-

sembly in an arbor press with the torsion bar hex opening facing upward. Support the outer edge of the control arm.

13. Use a brass drift, inserted into the hex opening, to press the pivot shaft out of the lower control arm. The inner shell of the bushing will remain on the shaft.

14. Cut the rubber portion of the bushing from the control arm. Remove the outer portion of the bushing with a chisel, or a hand press and a blunt drift.

CAUTION: *When using a chisel, do not cut into the control arm.*

Strut-to-crossmember bushing assembly

15. Remove the bushing inner shell from the pivot shaft by cutting it off if necessary.

Assembly and installation are done in the following order:

1. With its flange end first, fit a new bushing over the shaft. Press the bushing into the inner sleeve so that it seats on the shoulder of the shaft.

2. Press the shaft and the bushing on the control arm with a suitable drift and an arbor press.

3. Install the adjusting bolt and the swivel for the torsion bar. Place the strut in the lower control arm. Tighten the nut to 110 ft lbs (1968–69) or to 105 ft lbs (1970–73).

4. Fit the strut bushing inner retainer over the lower control arm shaft, then insert the shaft and the strut into the proper openings in the K-member.

5. Tighten the strut outer retainer and nut so they are finger-tight. Finger-tighten the lower control pivot shaft washer and nut as well.

6. Place the lower ball joint stud in the opening in the lower control arm. Tighten the nut to 115 ft lbs. Insert the cotter pin.

7. Install the brake support on the steering knuckle and finger-tighten its two upper bolts. Install the arm on the

steering knuckle, then finger-tighten the two lower bolts and nuts.

8. Tighten the nuts on the *upper* bolt to 55 ft lbs and nuts on the *lower* bolt to 120 ft lbs.

9. Check the tie-rod end seal for signs of damage and replace it if any are present. Attach the tie-rod end to the steering knuckle arm. Tighten its nut to 40 ft lbs. Slip the protector over the end seal and tie-rod end. Insert the cotter pin and install the stabilizer bar links (if so equipped).

10. Install the bolt, which attaches the shock absorber to the control arm. Finger-tighten it.

11. Install the torsion bar, as outlined, in Steps 7–11 (excluding Step 10) of the torsion bar installation procedure, above.

13. Install the wheel, tire, and brake assembly as described in Chapter 9.

14. Lower the car and make any necessary adjustments in suspension height Tighten the strut nut, at the K-member to 50–52 ft lbs. Insert the spring pin (if so equipped). Tighten the shock absorber bolt 50 ft lbs.

15. Tighten the lower control arm pivot shaft nut to 180–190 ft lbs.

16. Adjust the alignment to specification (see below).

1974–76

1. Turn the ignition lock to the "OFF" or "UNLOCKED" position.

2. Take the rebound bumper off.

3. Raise the car so that the front suspension is in full rebound (no load). Use *additional* jackstands under the front frame as a safety precaution.

4. Remove the wheel and tire assembly.

5. Remove the disc brake caliper, as detailed in Chapter 9.

CAUTION: *Do not hang the caliper by the brake line; safety wire it to the frame.*

6. Remove the wheel hub/brake disc assembly as detailed in Chapter 9.

7. Remove the nut, retainer, and bushing from the lower end of the shock.

8. Unfasten the two retaining bolts which secure the strut bar to the lower control arm.

9. Remove the stabilizer bar lower retaining bolt, retainer, and cushion from the lower control arm.

10. For assembly reference, measure the torsion bar anchor bolt depth into the lower control arm, then unload the torsion bar.

11. Perform the torsion bar removal procedure, detailed at the beginning of this chapter.

12. Using a puller, detach the ball joint from the steering knuckle arm.

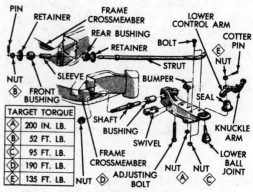

Lower control arm assembly—1974–76

13. Unfasten the lower control arm shaft nut and push the shaft out of the crossmember. If the shaft is frozen, strike its threaded end with a *soft* hammer.

14. Withdraw the lower control arm and shaft as assembly from under the car.

15. If necessary, perform Steps 11 through 15 of the "1968–73 Lower Control Arm Removal" procedure.

Assembly and installation are performed in the following order:

1. Perform Steps 1 through 3 of the "1968–73 Lower Control Arm Assembly and Installation" procedure. Skip the second part of Step 3.

2. Place the lower control arm/shaft assembly in the crossmember. Finger-tighten the control arm shaft nut.

3. Fit the end of the lower ball joint stud in the knuckle arm. Tighten its nut to 135 ft lbs.

4. Install the torsion bar as outlined at the beginning of this chapter. When loading the torsion bar, return the adjusting bolt to the depth in the control arm noted during removal.

5. Attach the strut bar to the lower control arm and tighten the bolts to 95 ft lbs. Secure the lower retaining bolt, re-

tainers, and cushions to the lower control arm; tighten the bolt to 30 ft lbs.

6. Install the brake dust shield.

7. Install the shock absorber insulator in the hole in the control arm. Install the bushing retainer, and nut on the shock bayonet. Tighten the nut to 35 ft lbs.

8. Install the brake disc/hub assembly and adjust wheel bearings. Install the caliper (see Chapter 9 for all these procedures).

9. Install the wheel and tire.

10. Lower the car to the ground. Check and adjust ride height.

11. Tighten the nut on the lower control arm shaft to 190 ft lbs.

12. Install the rebound bumper; tighten it to 200 in. lbs.

13. Have the wheel alignment checked and adjusted.

Upper Control Arm

REMOVAL AND INSTALLATION

1968–73

1. Follow Steps 1–3 of the upper ball joint removal procedure.

2. Remove the nuts, lockwashers, cams, and cam bolts which attach the upper control arm and bushings to the support brackets.

3. Withdraw the upper control arm by lifting it up and away from its support.

4. Remove the ball joint by using a suitable ball joint puller. The seal will come off together with the ball joint.

5. Firmly support the control arm; use a hammer and a drift to press the rear bushing out of the arm.

Assembly and installation of the upper control arm are performed in the following order:

1. Support the control arm squarely, and press the bushing into the control arm from outside, so that the tapered part of the bushing seats in the arm.

CAUTION: *Be sure that the control arm is supported squarely at the point where the bushing is being pressed in. Never use oil or grease to lubricate the bushing during installation.*

2. Install the ball joint to a *minimum* of 125 ft lbs. If a new arm is used, the ball joint will cut threads in it while it is being tightened.

3. Fit a new seal on the ball joint, being sure that it is fully seated in the ball joint housing.

4. Position the control arm, then install the cam bolts, cams, washers, and nuts in the reverse order from which they were removed.

5. Follow Steps 4–7 of the upper ball joint installation procedure.

1974–76

1. Turn the ignition lock to the "OFF" or "UNLOCKED" position.

2. Perform Steps 1 through 3 of the "Upper Ball Joint Removal" procedure.

3. Separate the ball joint from the knuckle.

4. Remove the rubber splash shield.

5. Unfasten the pivot shaft nuts.

6. Remove the upper control arm, complete with the ball joint and pivot bar, from the bracket.

7. Unfasten the nuts which secure the

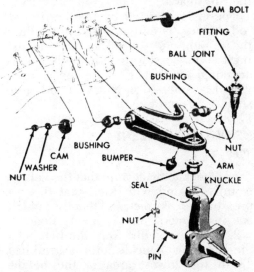

Upper control arm assembly—1968–73

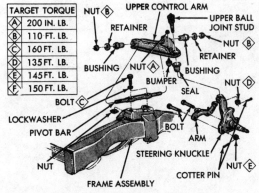

Upper control arm assembly—1974–76

TARGET TORQUE	
Ⓐ	200 IN. LB.
Ⓑ	110 FT. LB.
Ⓒ	160 FT. LB.
Ⓓ	135 FT. LB.
Ⓔ	145 FT. LB.
Ⓕ	150 FT. LB.

pivot bar to the upper control arm. Note their position and remove the retainers and bushings.

8. Use a puller to remove the ball joint from the upper control arm.

9. Press in new bushings if the old ones are excessively worn.

Installation of the upper control arm is performed in the reverse order of removal. For upper ball joint installation procedures, see the appropriate earlier section.

Front End Alignment

Because of the special equipment necessary to perform these adjustments, it is not possible to give exact procedures to accomplish these operations. In all cases, follow the alignment equipment manufacturer's recommendations. Set the front end to specifications. Some general notes on this operation follow.

Before attempting wheel alignment, the following points should be checked and corrected as necessary.

1. Front wheel bearing adjustment.

2. Check front wheel and tire for both radial and lateral run out.

3. Wheel and tire balance.

4. Ball joints and steering linkage pivots.

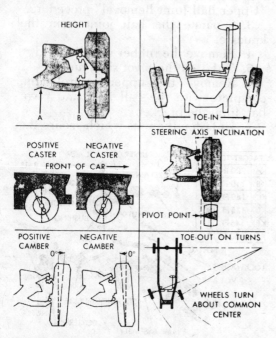

Six factors affecting wheel alignment

5. Check tire wear, size, and inflation for uniformity and proper specifications.

6. Shock absorbers.

7. Steering gear operation.

8. Rear springs.

9. Front suspension height.

CASTER AND CAMBER

Camber is the inward or outward tilt of the wheel. Camber is measured in degrees. A wheel that is tilted outward at the top has positive (+) camber. A wheel that is tilted inward at the top has negative (−) camber. Caster is the backward or forward tilt of the top of the steering knuckle. If the steering knuckle is tilted forward, it has negative caster. If the steering knuckle is tilted backward it has positive caster. Caster is also measured in degrees.

Caster and camber are adjusted by rotating the adjusting cams and locking them into position (1968–73). Caster should be as nearly equal as possible on both wheels. Camber sometimes differs slightly from side to side if the car is driven where the roads are crowned.

On 1974–76 models, caster and camber are adjusted by means of bolts on the lower control arm pivot bar.

TOE-IN

Toe-in is the amount by which the wheels are closer together at the front compared to the rear. This is measured in inches, between the edges of the wheel rims or between the centers of the tire treads.

Toe-in is adjusted by rotating the tie rod sleeves. The front wheels must be in the straight ahead position when adjusting toe-in. Be sure to turn both sleeves an equal amount. Be sure the steering wheel is centered in its travel and on its shaft.

HEIGHT ADJUSTMENT

1968–74

1. Check to make sure that the vehicle is fully loaded with fuel, that the tire pressures are correct, and that the vehicle is positioned on a level floor.

2. Clean road dirt from the bottom of the steering knuckle arm assemblies. Clean the lowest area of the height-adjusting blades directly below the

center of the lower control arm inner pivot assembly.

3. Bounce vehicle at least five times and release on a downward motion.

4. Check the distance from the bottom of one adjusting blade to the floor (refer to illustration measurement A) and from the lowest point of the steering knuckle arm at the centerline on the same side of the vehicle (measurement B). Be sure to measure only one side at a time.

5. The difference in measurement between A and B is the front suspension height.

6. Refer to the specifications and make adjustments as necessary. Do this by rotating the torsion bar adjusting bolt clockwise to increase the height and counterclockwise to decrease the height. After each adjustment, bounce the vehicle as was done previously before checking the height. Both sides must be measured even though only one side may have been adjusted. Be sure that height does not vary more than ⅛ in. from side to side.

1975–76

1. Jounce the car several times, releasing it on the downward motion.

2. Measure the distance between the lowest point of the lower control arm torsion bar anchor (at a point one inch forward of the rear face of the anchor) and the ground. This is measurement "A".

3. Compare measurement "A" with

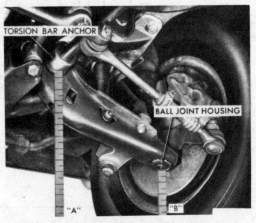

Ride height measurement—1968–74

the figure given in the "Ride Height" chart.

4. Adjust, if necessary, by rotating the torsion bar adjusting bolt clockwise to increase front end height, or rotating counterclockwise to decrease the height.

5. Check the adjustment on both sides. Maximum variation of front end height is ⅛ in.

Ride Height

Year	Model	Ride Height (in.)
1968–73	Fury, VIP	1⅜ ± ⅛
1974	Fury	1 ± ⅛
1975–76	Gran Fury	10⅛ ± ⅛

Wheel Alignment Specifications

Year	Model	CASTER Range (deg)	CASTER Pref Setting (deg)	CAMBER Range (deg)	CAMBER Pref Setting (deg)	Toe-in (in.)	Steering Axis Inclin.	WHEEL PIVOT RATIO (deg) Inner Wheel	WHEEL PIVOT RATIO (deg) Outer Wheel
1968–69	M.S.—Fury	0 to 1N	½N	①	①	3/32 to 5/32	7½	20	18.8
	P.S.—Fury	¼P to1¼P	¾P	①	①	3/32 to 5/32	7½	20	18.8

Wheel Alignment Specifications (cont.)

Year	Model	CASTER Range (deg)	CASTER Pref Setting (deg)	CAMBER Range (deg)	CAMBER Pref Setting (deg)	Toe-in (in.)	Steering Axis Inclin.	WHEEL PIVOT RATIO (deg) Inner Wheel	WHEEL PIVOT RATIO (deg) Outer Wheel
1970–72	M.S.—Fury	0 to 1N	½N	①	①	3/32 to 5/32	7½	20	18.8
	P.S.—Fury	¼ to 1¼P②	¾P②	①	①	3/32 to 5/32	7½	20	18.8
1973	P.S.—Fury	1/16N to 1 5/16P	⅝P	③	③	1/32 to 7/32	7½	20	18.8
1974	P.S.—Fury	½N to 1¾P	⅝P	④	④	1/16 to ¼	9	20	18.3
1975	P.S.—Gran Fury	1/16N to 1 5/16P	¾P	③	③	3/32 to 9/32	9	20	18.3
1976	P.S.—Gran Fury	½N to 1¾P	¾P	④	④	1/16 to ¼	9	20	18.3

M.S. Manual Steering
P.S. Power Steering
N Negative
P Positive

① Left—¼P to ¾P; ½P preferred
Right—0 to ½P; ¼P preferred
② 1970—0 to 1N; ½N preferred
③ Left—⅛P to ⅞P; ½P preferred
Right—⅛N to ⅝P; ¼P preferred
④ Left—0 to 1P; ½P preferred
Right—¼N to ¾P; ¼P preferred

REAR SUSPENSION

All Plymouth models utilize rear springs of the semi-elliptical leaf type. They are engineered to operate with little or no camber under conditions of small loads (including no load). Heavy-duty springs are offered as an option on all models. They increase the stability of the vehicle under conditions of heavy load. All vehicles equipped with leaf springs are constructed with zinc interleaves between the normal leaves. They have the purpose of reducing spring corrosion and lengthening spring life.

Shock absorbers used on Chrysler vehicles are not used to support vehicle load. Their sole purpose is to control ride motion.

Plymouth shock absorbers have a built-in fluid weep. This is usually evident only during cold weather. Consequently, a slight fluid weep is not reason to replace a shock absorber.

Shock Absorbers

REMOVAL AND INSTALLATION

1. Raise the car by a safe, suitable means, so that it is at convenient working height.
2. Place a floor stand under the rear axle to raise it and to take the load off from the shock absorbers.
3. Remove the nut and retainer which secure the shock to the spring plate mounting stud. Take the shock off

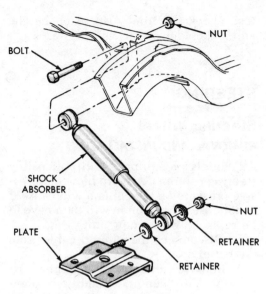

Rear shock absorber assembly

the stud. Remove any remaining washers from the stud.

4. Remove the nut and bolt from the upper shock absorber mount. Remove the shock from under the car.

5. Check the shock mounting bushings. If they show signs of damage or deterioration, they should be replaced.

6. Extend the shock fully by hand. Check for any sign of fluid running down the side of the unit

NOTE: *A slight amount of seepage is normal.*

7. Check to see that there is no air trapped inside the cylinder. Hold the shock in its normal *upright* operating position, while fully extending and compressing it. There should be no lost motion in either direction of travel.

8. If lost motion is apparent, continue to hold the shock upright, while fully extending it. Then invert the shock, while slowly compressing it.

CAUTION: *Never extend the unit while it is in an inverted position.*

9. Repeat Step 8 several times to expel all of the air which is trapped in the cylinder.

10. If the lost motion does not disappear after repeating the above, the shock is defective and must be replaced. Repeat Step 8 before installing the new shock absorber

NOTE: *New shocks may show greater resistance because the new seal generates more friction.*

Installation is performed in the following order:

1. Install the shock over the upper mounting stud and install the stud nut.

2. Fit the washer over the spring plate mounting stud and install the shock over that stud. Install the remaining washer and the nut. Do not tighten the nut fully.

3. Lower the car so that the full weight of it rests on the wheels. Tighten the upper nut to 70 ft lbs and the lower nut to 50 ft lbs.

Springs
REMOVAL AND INSTALLATION

1. Jack up the vehicle and remove the wheels. Position the jackstands under the axle in such a manner so as to relieve the weight on the rear springs.

2. Disconnect the rear shock absorbers at the bottom attaching bolts. Lower the axle assembly to allow the rear springs to hang free.

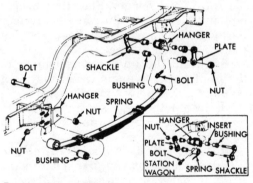

Rear spring—1968–73

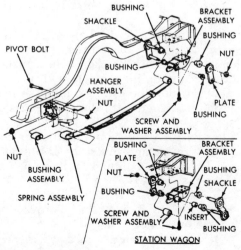

Rear spring—1974–76

3. Remove the U-bolt nuts and withdraw the bolts and spring plates. Remove the nuts securing the front spring hanger to the body mounting bracket.

NOTE: *On 1974–76 models, it will be necessary to install a special spring stretcher (tool #C-4211) before performing Step 3. Install the stretcher on the shackle and apply tension to the spring and remove the rear hanger assembly, then remove the front hanger. Do not try to remove the spring without the stretcher; it is preloaded and its sudden release could cause serious personal injury.*

4. Remove the rear spring hanger bolts and allow the spring to drop enough to allow the front spring hanger bolts to be removed.

5. Remove the front pivot bolt from the front spring hanger.

6. Remove the shackle nuts and remove the shackle from the rear spring.

7. To begin installation, assemble the shackle and bushings in the rear of the spring and hanger. Start the shackle bolt nuts. Do not lubricate the rubber bushings to ease installation. Do not tighten the bolt nut.

8. Install the front spring hanger to the front spring eye and insert the pivot bolt and nut. Do not tighten them.

9. Install the rear spring hanger to the body bracket and torque the bolts to 30 ft lbs.

NOTE: *On 1974–76 models install the hanger on the spring first. Then, with aid of the stretcher (tool #C-4211), install the rear hanger-to-frame bolts. Tighten the hanger bolts to 30 ft lbs and remove the stretcher.*

10. With the aid of a helper, raise the spring and insert the bolts in the spring hanger mounting bracket holes. Install the nuts and torque them to 30 ft lbs.

11. Position the axle assembly so it is correctly aligned with the spring center bolt.

12. Position the center bolt over the lower spring plate. Insert the U-bolt and nut. Torque the bolt to 45 ft lbs and connect the shock absorbers.

13. Lower the vehicle. Torque the pivot bolts to 125 ft lbs. Torque the shackle nuts to 30 ft lbs.

14. After this operation, drive the vehicle. Then, after completing the road test, check the front suspension height, and make adjustments as necessary.

STEERING

Steering Wheel

REMOVAL AND INSTALLATION

All models are equipped with collapsible steering columns. A sharp blow or excessive pressure on the column will cause it to collapse. The column will then have to be replaced with a new unit.

1. Disconnect the ground cable from the battery.

2. Remove the padded center assembly. This center assembly is often held on only by spring clips. However, on deluxe interiors it is held on by screws behind the arms of the wheel.

3. Remove the large center nut. Mark the steering wheel and steering shaft so that the wheel may be replaced in its original position.

4. Using a puller, pull the steering wheel from the steering shaft. It is possible to make a puller by drilling two holes in a piece of steel exactly the same distance apart as the two threaded holes on either side of the large nut. Drill another hole in the center of the piece the same diameter as the steering shaft. Find a bolt of a slightly smaller diameter than the steering shaft. Place the puller over the steering shaft and thread the two bolts into the holes in the wheel. Tighten the two bolts, and then tighten the center bolts to draw the wheel off the shaft.

5. Reverse the above procedure to install the wheel. When placing the wheel on the shaft, make sure the tires are straight ahead and the match marks are aligned.

Turn Signal Switch

REMOVAL AND INSTALLATION

1968–69

1. Disconnect the negative ground cable.

2. Disconnect the wiring connectors at the base of the steering column.

3. Remove the steering wheel as outlined above. Tie a string to the turn signal switch wires. Before proceeding further, see if the switch wiring will pull out of the switch without having to remove the wiring from the column.

4. Remove the turn signal lever which is screwed into the switch.

NOTE: *On models with cruise control, do not remove the turn signal lever, allow it to hang by the wire.*

5. Disconnect the switch wiring multiconnector at the base of the column jacket.

6. Unfasten the screws which secure the switch to the column. Remove the switch and its wiring from the column, leaving the string in the column as an installation aid.

Installation is performed in the reverse order of removal. Attach the switch wiring to the string which was left in the column and pull the wiring through the column with the string.

1970–76

1. Perform Steps 1 through 4 of the "1968–69 Turn Signal Switch Removal" procedure; it is unnecessary to tie a string to the switch wiring, however.

2. Remove the screws which attach the turn signal switch upper bearing retainer and remove the retainer.

3. If the column has a cover, remove it.

4. Unfasten the wire which holds the horn wire on its mounting stud.

5. Remove the nuts which attach the mounting bracket to the steering column.

6. Separate the wiring harnesss trough from the column by unfastening its screws. Remove the tape from the

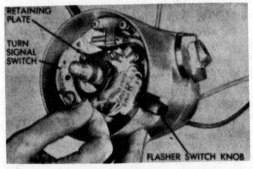

Removing the combination turn signal hazard warning switch—1971–76

harness and unfasten the harness multiconnector.

7. Pull the switch out of the column, while carefully guiding its wires through the column.

8. Work the connector through the column opening and completely remove the switch from the column.

Installation is the reverse of removal.

Ignition Switch (Column-Mounted)

REMOVAL AND INSTALLATION
1970–76 Standard Steering Column

1. Disconnect the negative battery cable. Remove the steering wheel.

2. Remove the screw that attaches the turn signal lever to the steering column. On cars with cruise control, do not completely remove the lever.

3. Remove the three screws that attach the upper bearing retainer to the turn signal switch.

4. Pull the turn signal switch as far upward as possible.

5. Using snap-ring pliers, remove the upper bearing housing snap-ring from the steering shaft.

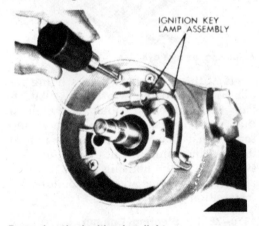

Removing the ignition key light

6. Remove the screw that attaches the ignition key light assembly to the upper bearing housing.

7. Using care not to damage any components, pry the upper bearing housing off the steering shaft by lifting upward on alternate sides of the bearing housing with screwdrivers.

8. Lift upward as far as possible on the steering shaft lock plate and place a

screwdriver or other object under it to hold it in the raised position. If this operation does not provide adequate working room under the lock plate, it will be necessary to press out the pin that attaches the lock plate to the steering shaft and remove the lock plate from the steering shaft. If the ignition switch is being replaced, the lock plate must be removed.

9. Using an offset screwdriver, remove the two screws that attach the lock lever guide plate to the steering column.

10. With the ignition lock cylinder in the "Lock" position and the ignition key removed, insert a stiff wire into the lock cylinder release hole in the steering column. Push in on the wire to release the spring-loaded lock retainer and pull the lock cylinder out of the steering column.

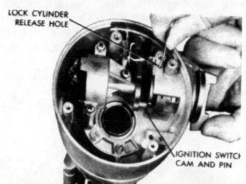

Removing the lock cylinder

11. If the ignition switch is being replaced, remove the two screws that attach the ignition key buzzer switch to the steering column and the three screws that attach the ignition switch to the steering column. Lift off the ignition switch out of the housing.

12. Reverse the above procedure for installation.

1970–76 Tilt Steering Column

1. Disconnect the negative battery cable.

2. Remove the steering wheel.

3. Remove the three attaching screws and remove the shaft lock cover.

4. Remove the screws that attach the tilt control lever and the turn signal lever to the steering column and then remove the levers.

5. Push in the hazard warning knob and unscrew the knob from the turn sig-

nal switch. Remove the ignition key lamp assembly.

6. Using a suitable tool, depress the lock plate to gain access to the lock plate retaining snap-ring. Remove the snap-ring from the steering shaft.

7. Remove the lock plate, cancelling cam, and spring.

8. Remove the three turn signal switch attaching screws, place the shift lever in the Low (L) position, and pull the switch and wires as far upward as possible.

9. With the ignition lock cylinder in the "Lock" position, insert a small screwdriver into the lock release slot in the housing cover.

10. Press down with the screwdriver to release the spring latch at the bottom of the slot and pull the lock cylinder from the housing. The following steps are for ignition switch replacement only.

11. Remove the three screws that attach the upper steering column housing to the steering column and remove the housing.

12. Install the column tilt control lever and move the column to the full "Up" position.

13. Insert a screwdriver into the slot in the spring retainer and press the retainer in approximately $3/16$ in. Turn the retainer approximately $1/8$ turn to the left until the ears align with the grooves in the housing. Remove the spring retainer, spring, and guide.

14. Push the steering shaft inward to enable removal of the inner race and seat. Remove the race and seat.

15. Make sure the ignition switch is in the "Lock" position, then remove the wire connector from the ignition switch and remove the screws that attach the ignition switch to the outside of the steering column.

16. Lift the ignition switch from the column and twist it to disengage the switch actuating rod from the rack. Remove the switch.

17. To install the ignition lock cylinder, insert the cylinder into the housing with the cylinder in the "Lock" position and the key *removed*.

18. Move the cylinder into the housing until it contacts the switch actuator. Move the switch actuator rod up and down to align the parts. When the parts

are aligned, the cylinder will move inward and lock into place. The following steps are for ignition switch installation only.

19. With the ignition switch in the "Lock" position, insert the actuating rod into the steering column.

20. Twist the switch and rod assembly as required to engage the actuating rod with the rack. Make sure the ignition lock cylinder is in the "Lock" position.

21. Install the ignition switch mounting screws but do not tighten them.

22. Move the ignition switch downward, away from the steering wheel, and tighten the switch mounting screws. Make sure that the ignition switch has not moved out of the lock detent.

23. Attach the switch wiring connector.

Power Steering Pump
REMOVAL AND INSTALLATION

Before beginning removal, take careful note of the exact hose routing. Hoses must be routed and installed in the exact same manner as they were removed. Read the entire procedure before beginning pump service.

1. Back off the pump mounting and locking bolts and remove the pump drive belt.

2. Disconnect all hoses at the pump.

3. Remove the pump bolts and the pump with the bracket.

4. To install a pump, place it in position and install the mounting bolts.

5. Install the pump drive belt and adjust it to specifications. Torque the mounting bolts to 25–30 ft lbs.

6. Connect the pressure and return hoses. On the 1.06 cu in. pump, install a new pressure hose O-ring.

7. Fill the pump with power steering fluid.

8. Turn on the motor and rotate the steering wheel from stop-to-stop at least ten times. This will bleed the system. Check the pump oil level and fill as required.

9. The torque of the pump-end hose fitting is 24 ft lbs. The gear end fitting torque is 160 in. lbs. Be certain that hoses are at least two inches from exhaust manifolds and are not kinked or twisted.

Tie-Rod Ends
INSPECTION

1. Raise the car under the lower control arm.

2. Make sure that the control arm ball joints are good and that the wheel bearings are adjusted. Grasp the tire on either side and move the tire from side-to-side. If excessive play is present (more than 1/8 in.), visually inspect the linkage while moving the tire, in order to determine exactly where the play is.

REMOVAL AND INSTALLATION

1. Loosen the tie-rod adjuster sleeve clamp nuts.

2. Remove the tie-rod stud nut and cotter pin.

3. If the outer tie-rod end is being removed, remove the ball joint stud from the steering knuckle. If the inner tie-rod end is being removed, remove the ball joint stud from the center link.

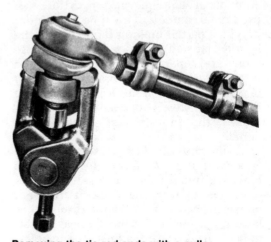

Removing the tie-rod ends with a puller

4. Unscrew the tie-rod end from the threaded sleeve. The threads may be left or right-hand threads. Count the number of turns required to remove the tie-rod end.

5. To install, reverse the above. Turn the tie-rod end in as many turns as was needed to remove it. This will give approximate correct alignment. Have the front end professionally realigned.

Brakes

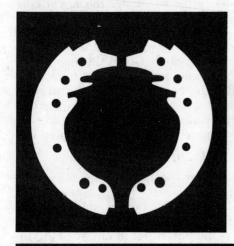

BRAKE SYSTEM

A dual or tandem master cylinder is used on all models. This type of cylinder provides partial braking if one half of the system fails. This is possible because there are separate fluid reservoirs and pistons for the front brakes and rear brakes. If there is a failure in the line to the rear brakes, you lose only the rear brakes, the front brakes remain functional.

All the drum brakes are self-adjusting. However, it is advisable to adjust the brakes manually since the self-adjusters often fail to work or work at different rates on each wheel. Front disc brakes have no provision for mechanical adjustment. Disc brakes adjust themselves. As the brake pad wears, the pad moves closer to the disc and stays in adjustment.

ADJUSTMENT

It is best to adjust all four wheels (drum brake models) at once but it is possible to adjust only the front or rear brakes. The adjustment must be equal on all wheels. The front adjustment is especially critical because a misadjusted brake will cause the car to pull or veer to one side. Remember, front disc brakes require no adjustment.

1. Lift up the front or rear of the car so that both wheels are free to turn. Take off the cover from the adjustment hole and insert an adjusting tool. Lift the handle of the tool upward (push downward 1968) and continue this until a slight drag is felt when the wheel is rotated.

2. Insert a piece of welding rod or a thin screwdriver, into the adjustment hole and disengage the adjusting lever from the starwheel.

CAUTION: *Do not bend the lever, or stretch its spring too far.*

3. Keep the lever disengaged while

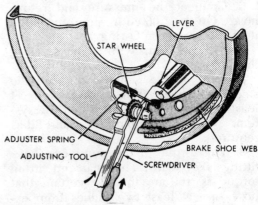

Manual brake shoe adjustment

backing off on the starwheel until there is no shoe drag when the wheel is rotated. Repeat the adjustment for the other front or rear wheel.

HYDRAULIC SYSTEM

Master Cylinder
REMOVAL AND INSTALLATION

1. Disconnect the brake lines from the master cylinder. Use a good wrench because the connections are very tight.

2. Remove the nuts that attach the master cylinder to the fire wall or power brake booster, if so equipped.

3. On models with standard brakes, disconnect the master cylinder pushrod from the brake pedal.

4. Slide the master cylinder straight out and off the firewall or brake booster.

5. Reverse the above procedure to install the master cylinder. When reconnecting the brake lines, start the fitting with your fingers and turn the fitting in several threads before using a wrench. This will prevent cross threading. If difficulty is encountered when threading the fittings, bend the brake line slightly so that the fitting enters the hole squarely. If a fluid leak occurs tighten the fitting, check for a damaged seat or tubing end, or look for a hair line crack in the tubing.

6. Bleed the brake system after installation is complete.

OVERHAUL

NOTE: *Although several types of master cylinders have been used, overhaul procedures are generally the same.*

1. Clean the outside of the cylinder. Remove the cover and drain the fluid.

2. Loosen the rear piston retainer screw in the flange below the piston and flip the retainer down to release the piston assembly for removal.

3. Remove the front piston. If the piston sticks in the cylinder, air pressure may be used to remove it. Always use new rubber cups.

4. Note the position of the rubber cups and springs and remove them from the pistons and from the bore.

5. Remove the tube seats, using an easy out or a screw threaded into the seat. Unless the seat is damaged it is not absolutely necessary to remove the seats.

6. Remove the residual pressure valves and springs found under the seats.

7. Clean the inside of the master cylinder with brake fluid or denatured alcohol.

8. Closely inspect the inside of the master cylinder. Polish the inside of the bore with crocus cloth. If there is rust or pit marks it will be necessary to use a hone. Discard the master cylinder if scores or pits cannot be eliminated by honing.

9. Do not reuse old rubber parts and be sure to use all the new parts supplied in the rebuilding kit.

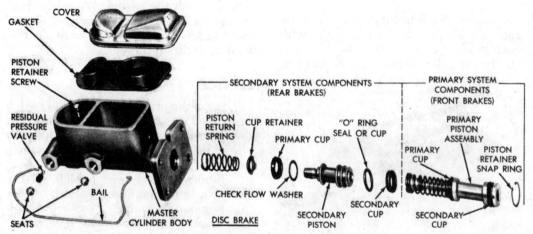

Typical master cylinder assembly

10. Before assembly, thoroughly lubricate all parts with clean brake fluid.

11. Replace the primary cup on the front end of the piston with the lip away from the piston.

12. Carefully slide the second seal cup over the rear of the piston and into the second land. The cup lip must face the front of the piston.

13. Slowly work the rear secondary cup over the piston and position it in the rear land. The lip must face to the rear.

14. Slide the retainer cup over the front piston stem with the beveled side facing away from the piston cup.

15. Replace the small end of the pressure spring into the retainer.

16. Position the assembly in the bore. Be sure the cups are not canted.

17. Slowly work the secondary cup over the back of the rear piston with the cup lip facing forward.

18. Position the spring retainer in the center of the rear piston assembly. It should be over the shoulder of the front piston. Position the piston assembly in the bore. Slowly work the cup lips into the bore, then seat the piston assembly.

19. Hold the pistons in the seated position. Insert the piston retaining screw with the gasket, and tighten it securely.

20. Replace the residual pressure valves and spring. Position them in the front outlet and install the tube seats.

BLEEDING THE MASTER CYLINDER

Before installing a reconditioned master cylinder, it will be necessary to bleed it.

1. Insert bleeding tubes into the tube seats and fill both brake reservoirs with brake fluid.

2. Insert a dowel pin into the depression in the piston and push in and release the piston. It will return under its own spring pressure. Repeat this operation until all of the air bubbles are expelled.

3. Remove the bleeding tubes and install the cover and the gasket.

4. Install the master cylinder on the car.

Pressure Differential Warning Valves

Since the introduction of dual master cylinders to the hydraulic brake system, a pressure differential warning signal has been added. This signal consists of a warning light on the dashboard activated by a differential pressure switch located below the master cylinder. The signal indicates a loss of fluid pressure in either the front or rear brakes, and should warn the driver that a hydraulic failure has occurred.

The pressure differential warning valve is a housing with the brake warning light switch mounted centrally on top. Directly below the switch is a bore containing a piston assembly. The piston assembly is located in the center of the bore and kept in that position by equal fluid pressure on either side. Fluid pressure is provided by two brake lines, one coming from the rear brake system and one from the front brakes. If a leak develops in either system (front or rear), fluid pressure to that side of the piston will decrease or stop causing the piston to move in that direction. The plunger on the end of the switch engages with the piston. When the piston moves off center, the plunger moves and triggers the switch to activate the warning light on the dash.

SERVICE

Service of the pressure differential switch is limited to its replacement. In order to replace the valve, proceed as follows:

1. Unfasten the hydraulic lines which run to the valve and plug them.

2. Remove the screws which secure the valve to its mounting bracket.

3. Remove the valve.

Installation is the reverse of removal. Remember to bleed the brake system

AIR BUBBLES

WOODEN DOWEL

BLEEDING TUBES

Bleeding the master cylinder

after installing the valve (see "Bleeding The Brakes").

Metering Valves

On some vehicles equipped with front disc brakes a metering valve is used. This valve is installed in the hydraulic line to the front brakes, and functions to delay pressure buildup to the front brakes on application. It provides balanced braking during mild stops. Its purpose is to reduce front brake pressure until rear brake pressure builds up adequately to overcome the rear brake shoe return springs. In this way disc brake pad life is extended because it prevents the front disc brakes from carrying all or most of the braking load at low operating line pressures.

The metering valve can be checked very simply. With the car stopped, gently apply the brakes. At about one inch of travel a very small change in pedal effort (like a small bump) will be felt if the valve is operating properly. Metering valves are not serviceable, and must be replaced if defective.

On some models the metering valve may be combined with the pressure differential warning switch.

REMOVAL AND INSTALLATION

1. Unfasten the hydraulic lines which run to the valve and plug them.
2. Remove the screws which secure the valve to its mounting bracket.
3. Remove the valve.

Installation is performed in the reverse order of removal. Remember to bleed the brake system after installing the valve (see "Bleeding The Brakes").

Proportioning Valves

On vehicles equipped with front disc and rear drum brakes a proportioning valve is an important part of the system. It is installed in the hydraulic line to the rear brakes. Its function is to maintain the correct proportion between line pressures to the front and rear brakes. It prevents early lock-up of rear brakes and provides balanced braking during hard stops. *No attempt at adjustment of this valve should be made, as adjustment is pre-set and tampering will result in uneven braking action.*

Combination line pressure warning/metering/proportioning valve

To assure correct installation when replacing the valve, the outlet to the rear brakes is stamped with the letter "R." Replacement is a simple job requiring no special instructions.

Beginning with 1972 models, Chrysler Corp. installed a combination valve on their front disc (rear drum) brake cars. This valve combines in one unit, a metering valve, a proportioning valve and a pressure differential warning valve. Mounted on top of the unit is an electrical terminal which connects to the brake warning light on the dash. This unit is not serviceable and must be replaced if faulty.

REMOVAL AND INSTALLATION

Service of the proportioning or combination valve is limited to its replacement. In order to replace the valve, proceed as follows:

1. Unfasten the hydraulic lines which run to the valve and plug them.
2. Remove the screws which secure the valve to its mounting bracket.
3. Remove the valve.

Installation is performed in the reverse order of removal. Remember to bleed the brake system after installing the valve (see the following procedure).

Bleeding the Brakes

1. Clean and fill the master cylinder with clean brake fluid. Keep the master cylinder filled during this procedure.

2. Jack up the car to allow access to the bleeder valves which are located on the inside of the brake plate.

3. Starting at the right rear wheel, attach a tube to the bleeder valve and insert the other end of the tube into a glass jar that has been partially filled with brake fluid (to check for air bubbles.).

4. Open the valve and have someone slowly push the brake pedal several times until the fluid coming out of the tube has no air bubbles.

5. Tighten the valve while steady pressure is being applied to the brake pedal.

6. Repeat the procedure for the rest of the wheels, first the left rear, then the right front, and finally the left front.

7. Throw away the fluid in the jar because it is full of microscopic air bubbles, then refill the master cylinder.

FRONT DISC BRAKES

Disc Brake Pads

REMOVAL AND INSTALLATION

To prevent paint damage from leaking brake fluid, remove some of the brake fluid from the master cylinder (do not reuse it) and keep the cylinder covered. Do not allow the cylinder fluid level to go too low or air will enter the hydraulic system.

1968

1. Raise the car and support it securely with jackstands.

2. Remove the wheel and tire.

3. Withdraw the brake pad anti-rattle spring.

4. Unfasten the caliper-to-steering knuckle bolts and slowly slide the caliper up, away from the disc.

5. Remove the brake pads one at a time through the caliper top opening.

6. Inspect the caliper for damaged or leaking seals or casting cracks.

7. To install new pads, insert the curved edge first (tabs, if any, should be up and position the steel plate against the pistons.

8. Spread the pads apart until the pistons are bottomed in their bores and slide the caliper assembly over the disc.

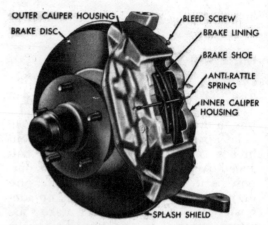

1968 disc brake caliper

9. Align the mounting holes and install the bolts; tighten to 80–90 ft lbs. Fit the anti-rattle springs over the pads.

10. Check the brake fluid level and pump the brake pedal to seat the linings against the disc. Replace the wheels and road-test the vehicle.

1969–76

1. Raise the vehicle on a hoist and remove front wheels.

2. Working on only one brake at a time, remove the caliper guide pins and positioners which attach caliper to adapter. Lift the caliper away from the disc.

3. Remove (and discard) the positioners and inner bushings from the guide pins, and the outboard bushings from the caliper.

4. Slide the disc pads out of the caliper, and carefully push the piston back into the bore.

Remove the caliper to gain access to the brake pads (shoes)—1969–72

5. Lubricate new outboard bushings and work them into position from the outboard side of the caliper.

6. Slide the new disc pads into position (outboard pad in the retaining spring) and carefully slide the caliper assembly over the rotor.

7. Lubricate and install new inner bushings in the caliper. Install new positioners on the guide pins with the open ends toward the outside.

8. Install the assembled guide pins from the inboard side and press in while threading pin into adapter. *Use extreme care to avoid crossing threads.* Tighten to 30–35 ft lbs. Be sure that the tabs of the positioners are over the machined surfaces of the caliper.

9. Check the brake fluid level and pump the brake pedal to seat the linings against the disc. Replace the wheels and road-test the car.

1973–76

1. Jack up the car and remove the wheel and tire.

2. Remove the caliper retaining clips and anti-rattle springs.

3. Remove the caliper from the disc by slowly sliding the caliper and brake pad assembly out and away from the disc.

4. Remove the outboard pad from the caliper by prying between the pad and the caliper fingers. Remove the inboard pad from the caliper support by the same method.

NOTE: *Safety-wire the caliper to the*

Removing the inboard pad (shoe)—1973–76

suspension while removing the inboard pad.

5. Push the pistons to the bottom of their bores. This may be done with a pair of large pliers or by placing a flat metal bar against the pistons and depressing the pistons with a steady force. This operation is much easier with the cover removed from the master cylinder.

6. Slide the new pads into the caliper and caliper support. The ears of the pad should rest on the bridges of the caliper.

7. Install the caliper on the disc and install the caliper retaining clips and anti-rattle springs. Pump the brake pedal until it is firm.

8. Check the fluid level in the master cylinder and add fluid as needed.

9. Install the wheel and tire.

10. Road-test the car. The car may pull to one side, but the pull should disappear shortly as the pads wear in.

Disc Brake Calipers
REMOVAL AND INSTALLATION
1968

1. Perform Steps 1, 2, and 4 of the "1968 Disc Brake Pad Removal" procedure. Skip Step 3.

2. Detach the brake line from the caliper and plug it.

3. Remove the pads, as outlined previously, if the caliper is to be overhauled.

Installation is performed in the reverse order of removal. See the disc pad installation procedure if the pads were removed. Check disc run-out (see below) prior to caliper installation. Caliper mounting bolts are tightened to 80–90 ft lbs. Bleed the system.

1969–76

1. Raise the car and support it securely with jackstands.

2. Remove the wheel and tire assembly from the car.

3. Detach the brake hose from the frame mounting bracket. Plug the brake tube to prevent fluid loss.

4. a. On 1969–72 models, remove the guide pins and positioners attaching the caliper to the adapter.

   b. On 1973–76 models, remove the screw, clip, and anti-rattle spring attaching the caliper to the adapter.

5. Slide the caliper assembly away

from the disc. Hold the outboard pad while doing this so that it can't fall out.

6. Remove the pads (see the appropriate earlier section) if the caliper is being overhauled.

Install the calipers as outlined under the appropriate pad replacement procedure. Connect the brake hose and bleed the brake system.

OVERHAUL

1968

1. Remove the caliper assembly from the car.

2. Open the caliper bleed screw and drain the caliper. Clean the outside of the caliper and place it in a vise with padded jaws.

3. Unfasten the four bridge bolts which secure the halves of the caliper. Separate the halves.

4. Remove the two crossover seals.

5. Use a small screwdriver to pry out the exposed end of the piston boot securing spring. Remove it from the groove to release the boot.

NOTE: *Keep the piston compressed during this operation.*

6. Remove the piston, seal, and dust boot from the caliper. Withdraw the piston return spring. Discard the old seal.

7. Remove the three other pistons in the same manner.

8. Clean all parts in denatured alcohol. Blow out all bores and passages with compressed air.

9. Replace any dust boots that are punctured or torn.

10. Inspect the bores for pitting or scoring. Light scratches may be cleaned with crocus cloth. Heavily scratched bores may be honed; do not enlarge the bore by more than 0.002 in. Replace the caliper if the bore can't be cleaned up without going over the above specification

NOTE: *Black stains in the bore are from piston seals and are harmless.*

11. After honing, flush the bore out with alcohol and dry with a clean, lint-free cloth. Repeat at least twice.

Caliper assembly is performed in the following order:

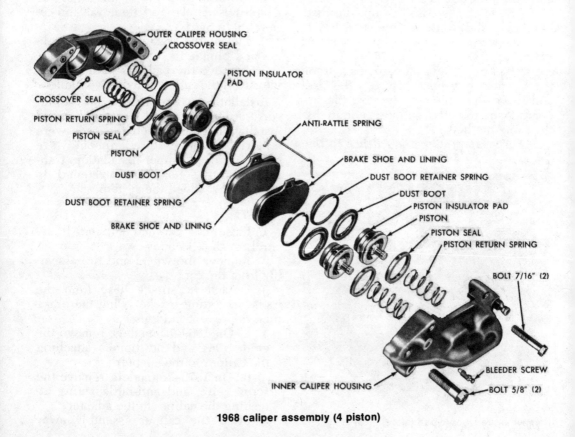

1968 caliper assembly (4 piston)

1. Coat the bores with brake fluid and then seat the piston return spring in the recess in the bottom of the bore.

2. Coat the new piston seal with brake fluid and work it down over the piston into its groove, using your fingers.

3. Install the boot on the piston so that the lip of the seat faces the piston pad.

4. Insert the piston in the bore, over the return spring, and depress it until it bottoms in the bore.

5. Work the lip of the boot into the groove around the diameter of the bore, using a small, blunt screwdriver. Use care not to puncture the boot or you will have to replace it.

6. Insert one end of the piston boot retaining spring into the groove and work it into place around the diameter of the bore, until it is fully seated.

CAUTION: *Be sure that the boot is fully secured by the spring and that the retainer is fully seated.*

7. Install the rest of the pistons by repeating Steps 1 through 6. Depress the pistons with your fingers to test for smooth operation.

8. With the outer caliper half clamped lightly in a vise, install new crossover passage seals in the recesses of the caliper mating surfaces.

9. Place the other caliper half over the one clamped in the vise and install the bridge bolts. Tighten them to the following specifications:

- $^7/_{16}$ in.—55 ft lbs.
- $^5/_8$ in.—150 ft lbs.

10. Install the brake pads, as previously detailed.

11. Fit the bleed screw and install the caliper on the car.

1969–72

1. Remove the caliper assembly from the car *without* disconnecting the hydraulic line (see above).

2. Support the caliper assembly on the upper control arm and surround it with shop towels to absorb any brake fluid. Slowly depress the brake pedal until the piston is pushed out of its bore.

CAUTION: *Do not use compressed air to force the piston from its bore; injury could result.*

3. Disconnect the brake line from the caliper and plug it to prevent fluid loss.

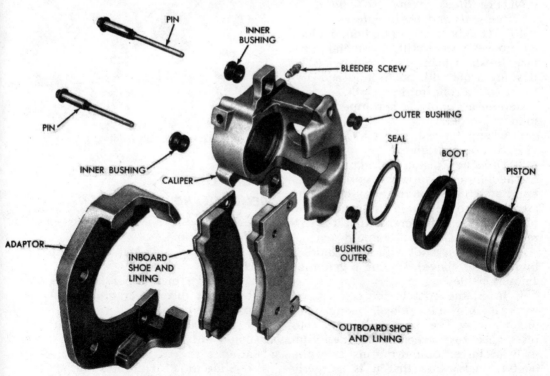

1969–72 caliper assembly (floating)

4. Mount the caliper in a soft-jawed vise and clamp lightly. Do not tighten the vise too much or the caliper will become distorted.

5. Work the dust boot out with your fingers.

6. Use a small pointed *wooden* or *plastic* stick to work the piston seal out of the groove in the bore. Discard the seal.

CAUTION: *Using a screwdriver or other metal tool could scratch the piston bore.*

7. Using the same wooden or plastic stick, press the outer bushings out of the housing. Discard the old bushings. Remove the inner bushings in the same manner. Discard them as well.

8. Clean all parts in denatured alcohol or brake fluid. Blow out all bores and passages with compressed air.

9. Inspect the piston and bore for scoring or pitting. Replace the piston if necessary. Bores with light scratches or corrosion may be cleaned with crocus cloth. Bores with deep scratches may be honed if you do not increase the bore diameter more than 0.002 in. Replace the housing if the bore must be enlarged beyond this.

NOTE: *Black stains are caused by piston seals and are harmless.*

10. If the bore had to be honed, clean its grooves with a stiff, non-metallic rotary brush. Clean the bore twice by flushing it out with brake fluid and drying it with a soft, lint-free cloth.

Caliper assembly is performed in the following order:

1. Clamp the caliper in a soft-jawed vise; do not overtighten.

2. Dip a new piston seal in brake fluid or the lubricant supplied with the rebuilding kit. Position the new seal in one area of its groove and gently work it into place with *clean* fingers, so that it is correctly seated. Do not use an old seal.

3. Coat a *new* boot with brake fluid or lubricant (as above), leaving a generous amount inside.

4. Insert the boot in the caliper and work it into the groove, using your fingers only. The boot will snap into place once it is correctly positioned. Run your forefinger around the inside of the boot to make sure that it is correctly seated.

5. Install the bleed screw in its hole and plug the fluid inlet on the caliper.

6. Coat the piston with brake fluid or lubricant. Spread the boot with your fingers and work the piston into the boot.

7. Depress the piston; this will force the boot into its groove on the piston. Remove the plug and bottom the piston in the bore.

CAUTION: *Apply uniform force to the piston or it will crack.*

8. Compress the flanges of new guide pin bushings and work into place by pressing *in* on the bushings with your fingertips, until they are seated. Make sure that the flanges cover the housing evenly on all sides.

9. Install the caliper on the car as previously outlined.

1973–76

The overhaul procedure for these calipers is identical to that for the 1969–73 calipers given above, except that there are no guide pin bushings to be removed or installed; omit the steps pertaining to them.

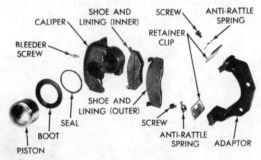

1973–76 caliper assembly (sliding)

Brake Disc
REMOVAL AND INSTALLATION

The brake disc and hub are removed at the same time.

1. Jack up the car and remove the tire and wheel.

2. Remove the caliper from the disc but do not disconnect the brake line. Support the caliper.

3. Remove the grease cup, cotter pin, locknut, thrust washer, and outer wheel bearing.

4. Slide the disc off the spindle.

5. Reverse the above to install the

disc. Tighten the wheel bearing nut to 90 in. lbs—1968–72 (or finger-tight) or to 240–300 in. lbs—1973–76.

INSPECTION

1. With the wheel removed, check to see that there is no grease or other foreign material on the disc. If the disc is badly scored replace the disc.

2. Measure the thickness of the disc with a micrometer at 12 points around the disc, 1 in. from the disc's edge. Any variation of more than 0.0005 in. means that the disc should be replaced.

ADJUST WHEEL BEARING TO ZERO LASH

BRAKING DISC

DIAL INDICATOR

PLUNGER CONTACTING DISC APPROXIMATELY 1 INCH FROM OUTER EDGE OF DISC

Checking disc run-out

3. Using a dial indicator, check the disc run out on both sides. Run-out should be no greater than the figures specified below. If the run-out exceeds those figures the disc should be replaced. Make sure that the wheel bearing is tight when making this measurement.

- 1968—0.0050 in.
- 1969–72—0.0025 in.
- 1973–76—0.0040 in.

Wheel Bearings

REMOVAL AND INSTALLATION

1. Jack up the car and remove the wheel and tire.

2. Remove the caliper from the disc but not from the car.

3. Remove the brake disc from the car.

4. When removing the inner wheel bearing, remove the grease seal. This seal must be replaced with a new seal. Do not reuse an old seal.

5. Remove the wheel bearing inner race and bearing with your fingers.

6. To remove the outer race from the hub, drive the race out of the hub with a long punch and hammer. There are two notches in the shoulder against which the race is seated, to provide access for the punch.

7. Installation is the reverse of removal. The outer race must be driven in place with a non-metallic rod. Force wheel bearing grease between all the rollers of the wheel bearing and fill the hub cavity with grease.

ADJUSTMENT

1. Tighten the adjusting nut to 90 in. lbs—1968–72 and to 240–300 in. lbs—1973–76, while rotating the wheel with a wrench.

2. Back off the adjusting nut to completely release any bearing preload.

3. Then finger-tighten the adjusting nut and install the locknut with a cotter pin. The resulting adjustment should yield no more than 0.003 in. of end-play.

FRONT DRUM BRAKES

Brake Drums

REMOVAL AND INSTALLATION

1. Jack up the car and remove the wheel and tire.

2. Remove the wheel bearing cover, cotter pin, lock and adjusting nut, and wheel bearing.

3. Withdraw the hub, drum, and bearing assembly from the spindle. The assembly may not slip easily off the spindle either because the brake is adjusted too tight or because there is a ring of rust at the inside edge of the drum. In either case the brake adjustment must be backed off.

4. To back off on the brake adjustment, remove the rear plug from the backing plate and loosen the starwheel until the drum will slip from the spindle.

5. Reverse the above steps to install. Adjust the brakes and wheel bearings.

INSPECTION

1. Drum run-out (out-of-round) and diameter should be measured. Drum diameter cannot exceed specification by more than 0.002 in. and run-out cannot exceed 0.006 in. Do not reface a drum

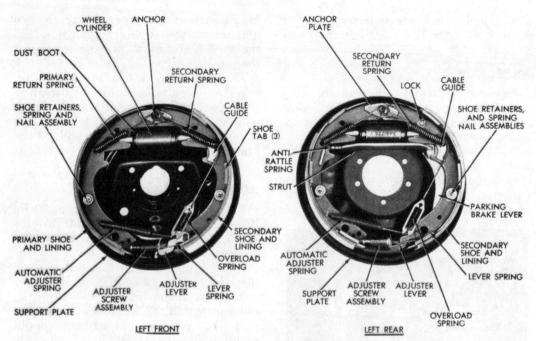

11 in. drum brake assembly

more than 0.060 inches over its standard diameter.

2. Check the drum for large cracks and scores. Replace the drum if necessary.

3. If the brake linings are wearing more on one edge than the other then the drum may be "bell" shaped and will have to be replaced or resurfaced.

Brake Shoes

REMOVAL AND INSTALLATION

NOTE: *Leave the brake shoes installed on one side, so that you can refer to them during assembly.*

Remove the wheel and brake drum and proceed as follows:

1. Remove the shoe return springs. Detach the adjusting cable eye from the anchor and unhook the other end from the lever. Withdraw the cable, overload spring, and anchor plate.

2. Detach the adjusting lever from the spring, and separate the spring from the pivot. Remove the spring completely from the secondary shoe web and unfasten it from the primary shoe web.

3. Remove the retainer springs and nails from the shoe. Extract both shoes from the pushrods, and lift them out. Withdraw the star wheel assembly from the shoes.

Install the brakes in the following order:

1. Lightly lubricate the six shoe tab contact areas on the support plate with Lubriplate®. Match both the primary and secondary brake shoes with each other.

2. Before installation in the car, fit the starwheel assembly between the shoes with the starwheel next to the secondary shoe. The left starwheel is plated and its adjustment stud is stamped with an "L."

3. Place the assembly on the support plate while attaching the shoe ends to the pushrods.

4. Install the shoe retaining nails, springs, and retainers. Place the anchor plate over the anchor.

5. Place the adjustment cable eye over the anchor, so that it rests against the anchor plate. Attach the primary shoe return spring to the shoe web and fit the other end over the anchor.

6. Place the cable guide in the secondary shoe web and fit the end over the anchor. Hold this in position while engaging the secondary shoe return spring, through the guide and into the web. Put its other end over the anchor. Make sure that the cable guide stays flat against the web and that the secondary shoe return spring overlaps the primary shoe return spring.

7. Squeeze the ends of the spring

loops until they are parallel and around the anchor.

8. The adjustment cable should be threaded over the guide and the end of the overload spring should be hooked in to the lever. The eye of the adjuster cable must be tight against the anchor.

9. Install the brake drum, wheel bearing, and wheel and tire. Adjust the brakes.

Wheel Cylinders

OVERHAUL

When the brake drums are removed, carry out an inspection of the wheel cylinder boots for cuts, tears, cracks, or leaks. If any of these are present, the wheel cylinder should have a complete overhaul performed.

NOTE: *Preservative fluid is used during assembly; its presence in small quantities does not indicate a leak.*

To remove and overhaul the wheel cylinders, proceed in the following manner:

1. Remove the brake shoes (see above) and check them. Replace them if they are soaked with grease or brake fluid.

2. Detach the brake hose.

3. Unfasten the wheel cylinder attachment bolts and slide the wheel cylinder off its support.

4. Pry the boots off from either end of the wheel cylinder and withdraw the push rods. Push in on one of the pistons, to force out the other piston, its cup, the spring, the spring cup, and the piston, itself.

5. Wash the pistons, the wheel cylinder housing, and the spring in fresh brake fluid, or in denatured alcohol, and dry them off using compressed air.

CAUTION: *Do not use a rag to dry them since the lint from it will stick to the surfaces.*

Inspect the cylinder bore wall for signs of pitting, scoring, etc. If it is badly scored or pitted, the entire cylinder should be replaced. Light scratches or corrosion should be cleaned up with crocus cloth.

NOTE: *Disregard the black stains from the piston cups that appear on the cylinder wall; they will do no damage.*

1. Dip the pistons and the cups in clean brake fluid. Replace the boots with new ones, if they show wear or deterioration. Coat the wall of the cylinder bore with clean brake fluid.

2. Place the spring in the cylinder bore. Position the cups in either end of the cylinder with the open end of the cups facing inward (toward each other).

3. Place the pistons in either end of the cylinder bore with the recessed ends facing outward (away from each other).

4. Fit the boots over the ends of the cylinder and push down until each boot

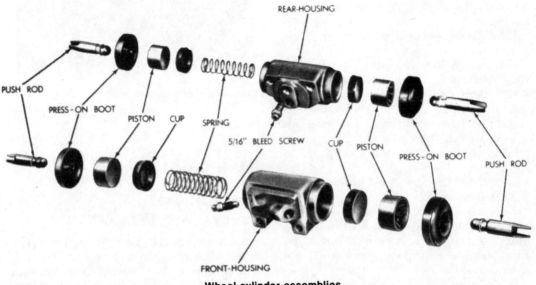

Wheel cylinder assemblies

is seated, being careful not to damage either boot.

5. Position the wheel cylinder on its support and torque the mounting screws to 220 in. lbs.

6. Attach the brake tube to the wheel cylinder and torque it to 115 in. lbs. Install the brake hose on the frame bracket. Connect the brake line to the hose, using a torque setting of 115 in. lbs also. Fit the end of the brake hose through the end of the stand-off. Attach the jumper tube to the brake hose and attach the hose to the stand-off. Torque the jumper tube to 195 in. lbs.

Wheel Bearings

REMOVAL AND INSTALLATION

The procedure for removing, installing, and adjusting the bearings on disc-brake-equipped cars appears earlier under "Brake Disc—Removal and Installation".

Drum brake wheel bearings are removed in the following manner:

1. Lift the car so that both of the wheels are clear of the ground.

2. Take off the wheel disc, grease cup, cotter pin, lock, and the bearing adjustment nut. Withdraw the thrust washer and the bearing cone.

3. Remove the wheel, hub, and brake drum from the spindle.

4. Use a ¾ in. *non-metallic* rod to drift out the inner oil seal and the bearing.

Clean the hub/drum assembly and the bearings by using mineral spirits or kerosine.

CAUTION: *Never dry the bearings by air-spinning them.*

Check the bearing cups for signs of wear or damage. If these are present, use a soft steel drift placed in the slots of the hub. The area into which the cup fits should be smooth and free of scoring. All cones, rollers, etc., should be free of scoring, chipping, or other damage, as well.

Installation is performed in the following manner:

1. If it was found necessary to install a new bearing cup, drive it flush with the hub, using a brass block and hammer. Be sure that the cup is seated against the shoulder of the hub.

2. Fill the grease cavity with wheel

bearing grease until the grease is even with the inner diameter of the bearing cups.

CAUTION: *Do not mix different brands of grease; not all of them are compatible. Because of this, new grease should never be added to that which is already in the hub.*

The grease should be forced into the bearing rollers or packed in with a bearing packer.

3. Insert the inner cone and put a new oil seal in position with its lip facing inward. Lightly tap the oil seal in place, using a wood block and a hammer so that it is flush with the hub. Use care not to damage the oil seal flange.

4. Clean the spindle. Coat its polished surfaces with a small amount of bearing grease. Fit the wheel and drum assembly over the spindle.

5. Position the outer bearing cone, the thrust washer, and the adjustment nut on the spindle. Adjust the preload as previously detailed.

Adjustment

Adjustment is the same as in the "Disc Brake Wheel Bearing Adjustment" section.

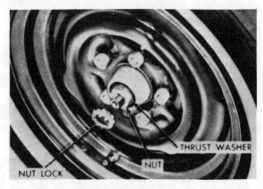

Front wheel bearing assembly

REAR DRUM BRAKES

Brake Drums

REMOVAL AND INSTALLATION

1. Remove the rear plug from the brake adjusting access hole.

2. Slide a thin screwdriver through the hole and position the adjusting lever

away from the adjusting notches on the star wheel.

3. Insert an adjusting tool into the brake adjusting hole and engage the starwheel. Pry downward with the tool to back off the brake adjustment.

4. Remove the rear wheel and tire. Remove the clips (if so equipped) from the wheel studs and discard the clips.

5. Remove the drum from the axle. The drum simply slips from the axle leaving the wheel studs in place in the axle. However, the drum will sometimes be rusted in place. To break the rust, strike the drum sharply several times with a heavy hammer on the corner of the drum. Strike the drum in several places around its circumference. Do not strike the drum on the edge of the open side as this will crack the drum.

6. Installation is the reverse of the above procedures.

INSPECTION

Inspection procedures are the same as the front drum inspection procedures.

Brake Shoes

REMOVAL AND INSTALLATION

NOTE: *Leave the brake shoe assembly installed on the side that you are not working on, so that you may refer to it during assembly.*

Except for the following steps, rear brake shoe removal is identical to that for the front brake shoes.

1. With the anchor ends of both shoes spread apart, remove the parking brake lever strut, as well as the anti-rattle spring.

2. Detach the parking brake cable from the parking brake lever.

The installation procedure for the rear brake shoes is different than that for the front shoes.

1. Put a thin film of lubricant at the six shoe tab contact areas on the support plate.

2. Lubricate the pivot on the inner side of the secondary shoe web, and install the parking brake lever on it. Fasten the lever with its washer and horseshoe clip.

3. Connect the parking brake cable to the lever. Slip the secondary shoe next to the support plate, while engaging the shoe web with the pushrod, and push it against the anchor.

4. Position the parking brake strut behind the hub and slide it into the slot in the lever. Fit the anti-rattle spring over the free end of the strut.

5. Position the primary shoe, engage it in the pushrod and with the free end of the parking brake strut. Place the anchor plate over the anchor and fit the eye of the adjustment cable over the anchor. Connect the primary shoe return spring to its web and fit its other end over the anchor.

6. Place the cable guide in the secondary shoe web. Hold it in this position while engaging the secondary shoe return spring, which goes through the guide and into the web. Put its other end over the anchor.

NOTE: *See that the cable guide stays flat against the web, and that the secondary shoe return spring overlaps that of the primary.*

Squeeze the spring loops around the anchor, with pliers, until they are parallel.

7. Place the starwheel assembly between the shoes, with the starwheel assembly adjacent to the secondary shoe. The left rear starwheel is plated and marked with an "'L."

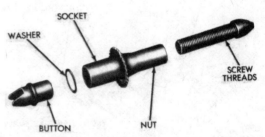

Starwheel adjuster assembly

8. Place the adjustment lever spring over the pivot pin on the shoe web and fit the lever under the spring, but over the pin. To lock the lever, push it toward the rear.

9. Install the shoe retaining nails, retainers, and spring. Thread the adjusting cable over the guide. Hook the end of the overload spring in the adjustment lever, making sure that the cable remains tight against the anchor and is aligned with the guide.

10. Install the brake drum and adjust

the brakes. Adjustment is the same as front drum brakes

Wheel Cylinders

OVERHAUL

The wheel cylinders found on the rear drum brakes are identical with those used on models equipped with front drum brakes. Service procedures for rear wheels may, therefore, be found with those detailed previously in the "Front Drum Brakes" section.

PARKING BRAKE

Cable

REMOVAL AND INSTALLATION

1. Jack up the car and remove the rear wheels.
2. Disconnect the brake cable from the equalizer.
3. Remove the retaining clip from the brake cable bracket.
4. Remove the brake drum from the rear axle.

Removing the parking brake cable from the support

5. Remove the brake shoe retaining springs and return springs.
6. Remove the brake shoe strut and spring from the brake support and disconnect the brake cable from the operating arm.
7. Compress the retainers on the end of the brake cable housing and remove the cable from the support.
8. Installation is the reverse of removal.

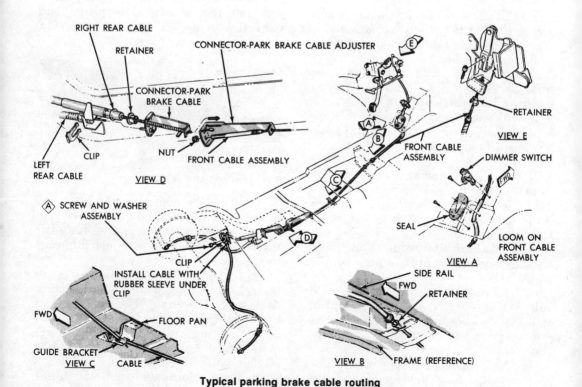

Typical parking brake cable routing

ADJUSTMENT

The brakes must be properly adjusted before the parking brake is adjusted.

1. Release the parking brake lever and loosen the cable adjusting nut.
2. Tighten the cable adjusting nut until a light drag is felt while rotating the wheel. Loosen the cable adjusting nut until both rear wheels can be rotated freely, then back off the cable adjusting nut two full turns.
3. Apply the parking brake several times and test to see that the rear wheels rotate freely.

Brake Specifications

Year	Model	MASTER CYLINDER Disc	Drum	WHEEL CYLINDER Front Disc	Drum	Rear	BRAKE DISC OR DRUM DIAMETER Front Disc	Drum	Rear
1968–69	Fury, VIP	1.125	1.0	2.375 ①	1.125	0.9375	11.75	11.0	11.0
1970	Fury	1.125	1.0	2.750	1.125	0.9375	11.75	11.0	11.0
1971–72	Fury	1.00	1.0	2.750	1.187	0.9375	11.75	11.0	11.0
1973–75	Fury, Gran Fury	1.03	——	2.750 ②	——	0.9375	11.75	——	11.0
1976	Gran Fury	1.03	——	3.100	——	0.9375	11.62	——	11.0

① 1969—2.750 in.
② 3.100 beginning 1974
—— Not applicable

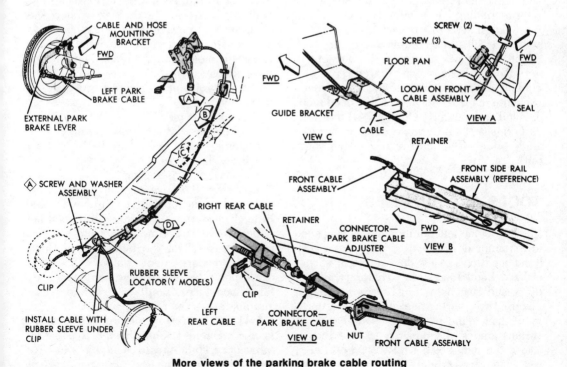

More views of the parking brake cable routing

Body

10

You can repair most minor auto body damage yourself. Minor damage usually falls into one of several categories: (1) small scratches and dings in the paint that can be repaired without the use of body filler, (2) deep scratches and dents that require body filler, but do not require pulling, or hammering metal back into shape and (3) rust-out repairs. The repair sequences illustrated in this chapter are typical of these types of repairs. If you want to get involved in more complicated repairs including pulling or hammering sheet metal back into shape, you will probably need more detailed instructions. Chilton's *Minor Auto Body Repair, 2nd Edition* is a comprehensive guide to repairing auto body damage yourself.

TOOLS AND SUPPLIES

The list of tools and equipment you may need to fix minor body damage ranges from very basic hand tools to a wide assortment of specialized body tools. Most minor scratches, dings and rust holes can be fixed using an electric drill, wire wheel or grinder attachment, half-round plastic file, sanding block, various grades of sandpaper (#36, which is coarse through #600, which is fine) in both wet and dry types, auto body plastic,

primer, touch-up paint, spreaders, newspaper and masking tape.

Most manufacturers of auto body repair products began supplying materials to professionals. Their knowledge of the best, most-used products has been translated into body repair kits for the do-it-yourselfer. Kits are available from a number of manufacturers and contain the necessary materials in the required amounts for the repair identified on the package.

Kits are available for a wide variety of uses, including:

- Rusted out metal
- All purpose kit for dents and holes
- Dents and deep scratches
- Fiberglass repair kit
- Epoxy kit for restyling.

Kits offer the advantage of buying what you need for the job. There is little waste and little chance of materials going bad from not being used. The same manufacturers also merchandise all of the individual products used—spreaders, dent pullers, fiberglass cloth, polyester resin, cream hardener, body filler, body files, sandpaper, sanding discs and holders, primer, spray paint, etc.

CAUTION: *Most of the products you will be using contain harmful chemicals, so be extremely careful. Always read the complete label before opening the containers. When*

you put them away for future use, be sure they are out of children's reach!

Most auto body repair kits contain all the materials you need to do the job right in the kit. So, if you have a small rust spot or dent you want to fix, check the contents of the kit before you run out and buy any additional tools.

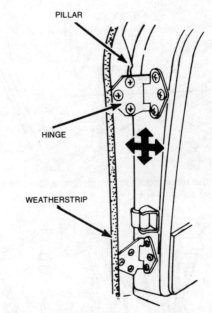

Door hinge adjustment

ALIGNING BODY PANELS

Doors

There are several methods of adjusting doors. Your vehicle will probably use one of those illustrated.

Whenever a door is removed and is to be reinstalled, you should matchmark the position of the hinges on the door pillars. The holes of the hinges and/or the hinge attaching points are usually oversize to permit alignment of doors. The striker plate is also moveable, through oversize holes, permitting up-and-down, in-and-out and fore-and-aft movement. Fore-and-aft movement is made by adding or subtracting shims from behind the striker and pillar post. The striker should be adjusted so that the door closes fully and remains closed, yet enters the lock freely.

DOOR HINGES

Don't try to cover up poor door adjustment with a striker plate adjustment. The gap on each side of the door should be equal and uniform and there should be no metal-to-metal contact as the door is opened or closed.

1. Determine which hinge bolts must be loosened to move the door in the desired direction.

2. Loosen the hinge bolt(s) just enough to allow the door to be moved with a padded pry bar.

3. Move the door a small amount and check the fit, after tightening the bolts. Be sure that there is no bind or interference with adjacent panels.

4. Repeat this until the door is properly positioned, and tighten all the bolts securely.

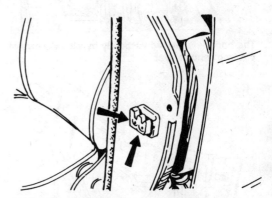

Move the door striker as indicated by arrows

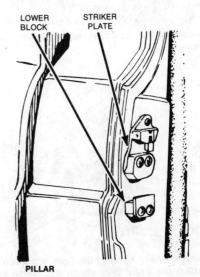

Striker plate and lower block

Hood, Trunk or Tailgate

As with doors, the outline of hinges should be scribed before removal. The hood and trunk can be aligned by loosening the hinge bolts in their slotted mounting holes and moving the hood or trunk lid as necessary.

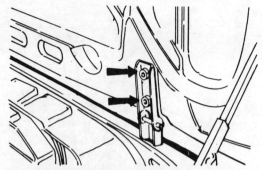

Loosen the hinge boots to permit fore-and-aft and horizontal adjustment

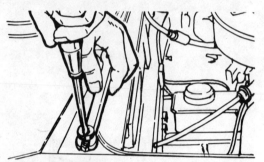

The hood is adjusted vertically by stop-screws at the front and/or rear

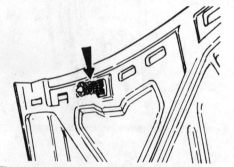

The hood pin can be adjusted for proper lock engagement

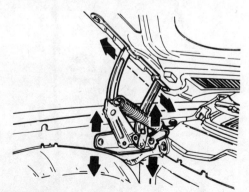

The height of the hood at the rear is adjusted by loosening the bolts that attach the hinge to the body and moving the hood up or down

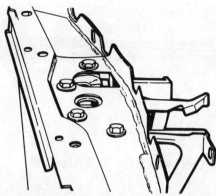

The base of the hood lock can also be re-positioned slightly to give more positive lock engagement

The hood and trunk have adjustable catch locations to regulate lock engagement. Bumpers at the front and/or rear of the hood provide a vertical adjustment and the hood lockpin can be adjusted for proper engagement.

The tailgate on the station wagon can be adjusted by loosening the hinge bolts in their slotted mounting holes and moving the tailgate on its hinges. The latchplate and latch striker at the bottom of the tailgate opening can be adjusted to stop rattle. An adjustable bumper is located on each side.

RUST, UNDERCOATING, AND RUSTPROOFING

Rust

Rust is an electrochemical process. It works on ferrous metals (iron and steel) from the inside out due to exposure of unprotected surfaces to air and moisture. The possibility of rust exists practically nationwide—anywhere humidity, industrial pollution or chemical salts are present, rust can form. In coastal areas, the problem is high humidity and salt air; in snowy areas, the problem is chemical salt (de-icer) used to keep the roads clear, and in industrial areas, sulphur dioxide is present in the air from industrial pollution and is changed to sulphuric acid when it rains. The rusting process is accelerated by high temperatures, especially in snowy areas, when vehicles are driven over slushy roads and then left overnight in a heated garage.

Automotive styling also can be a contributor to rust formation. Spot welding of panels creates small pockets that trap moisture and form an environment for rust formation. Fortunately, auto manufacturers have been working hard to increase the corrosion protection of their products. Galvanized sheet metal enjoys much wider use, along with the increased use of plastic and various rust retardant coatings. Manufacturers are also designing out areas in the body where rust-forming moisture can collect.

To prevent rust, you must stop it before it gets started. On new vehicles, there are two ways to accomplish this.

First, the car or truck should be treated with a commercial rustproofing compound. There are many different brands of franchised rustproofers, but most processes involve spraying a waxy "self-healing" compound under the chassis, inside rocker panels, inside doors and fender liners and similar places where rust is likely to form. Prices for a quality rustproofing job range from $100–$250, depending on the area, the brand name and the size of the vehicle.

Ideally, the vehicle should be rustproofed as soon as possible following the purchase. The surfaces of the car or truck have begun to oxidize and deteriorate during shipping. In addition, the car may have sat on a dealer's lot or on a lot at the factory, and once the rust has progressed past the stage of light, powdery surface oxidation rustproofing is not likely to be worthwhile. Professional rustproofers feel that once rust has formed, rustproofing will simply seal in moisture already present. Most franchised rustproofing operations offer a 3–5 year warranty against rust-through, but will not support that warranty if the rustproofing is not applied within three months of the date of manufacture.

Undercoating should not be mistaken for rustproofing. Undercoating is a black, tar-like substance that is applied to the underside of a vehicle. Its basic function is to deaden noises that are transmitted from under the car. It simply cannot get into the crevices and seams where moisture tends to collect. In fact, it may clog up drainage holes and ventilation passages. Some undercoatings also tend to crack or peel with age and only create more moisture and corrosion attracting pockets.

The second thing you should do immediately after purchasing the car is apply a paint sealant. A sealant is a petroleum based product marketed under a wide variety of brand names. It has the same protective properties as a good wax, but bonds to the paint with a chemically inert layer that seals it from the air. If air can't get at the surface, oxidation cannot start.

The paint sealant kit consists of a base coat and a conditioning coat that should be applied every 6–8 months, depending on the manufacturer. The base coat must be applied before waxing, or the wax must first be removed.

Third, keep a garden hose handy for your car in winter. Use it a few times on nice days during the winter for underneath areas, and it will pay big dividends when spring arrives. Spraying under the fenders and other areas which even car washes don't reach will help remove road salt, dirt and other build-ups which help breed rust. Adjust the nozzle to a high-force spray. An old brush will help break up residue, permitting it to be washed away more easily.

It's a somewhat messy job, but worth it in the long run because rust often starts in those hidden areas.

At the same time, wash grime off the door sills and, more importantly, the under portions of the doors, plus the tailgate if you have a station wagon or truck. Applying a coat of wax to those areas at least once before and once during winter will help fend off rust.

When applying the wax to the under parts of the doors, you will note small drain holes. These holes often are plugged with undercoating or dirt. Make sure they are cleaned out to prevent water build-up inside the doors. A small punch or penknife will do the job.

Water from the high-pressure sprays in car washes sometimes can get into the housings for parking and taillights, so take a close look. If they contain water merely loosen the retaining screws and the water should run out.

Repairing Scratches and Small Dents

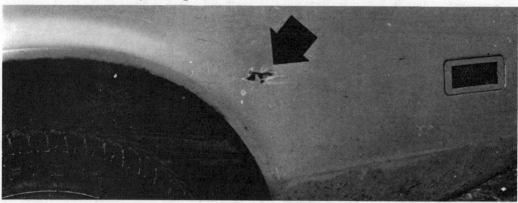

Step 1. This dent (arrow) is typical of a deep scratch or minor dent. If deep enough, the dent or scratch can be pulled out or hammered out from behind. In this case no straightening is necessary

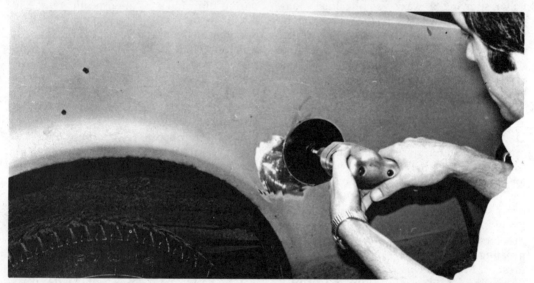

Step 2. Using an 80-grit grinding disc on an electric drill grind the paint from the surrounding area down to bare metal. This will provide a rough surface for the body filler to grab

Step 3. The area should look like this when you're finished grinding

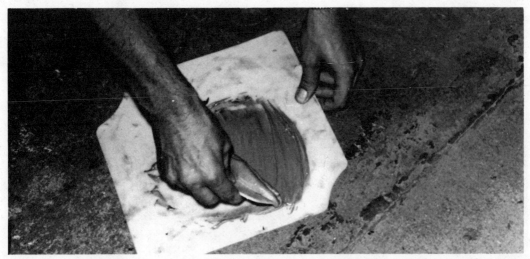

Step 4. Mix the body filler and cream hardener according to the directions

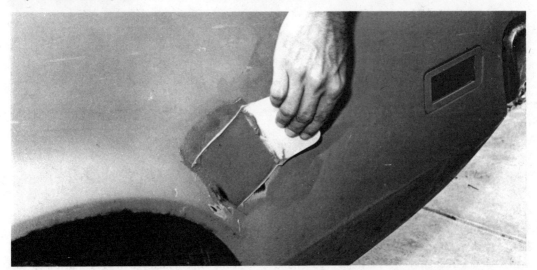

Step 5. Spread the body filler evenly over the entire area. Be sure to cover the area completely

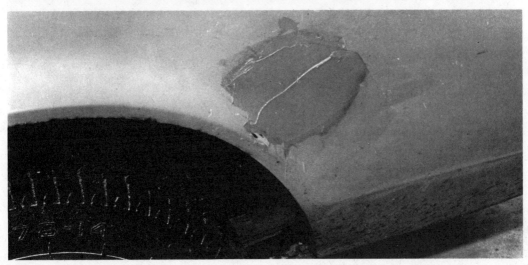

Step 6. Let the body filler dry until the surface can just be scratched with your fingernail

Step 7. Knock the high spots from the body filler with a body file

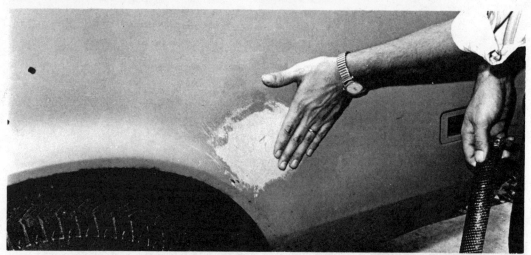

Step 8. Check frequently with the palm of your hand for high and low spots. If you wind up with low spots, you may have to apply another layer of filler

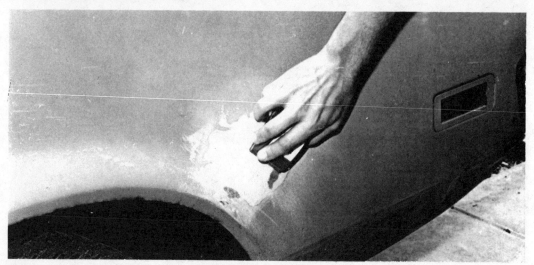

Step 9. Block sand the entire area with 320 grit paper

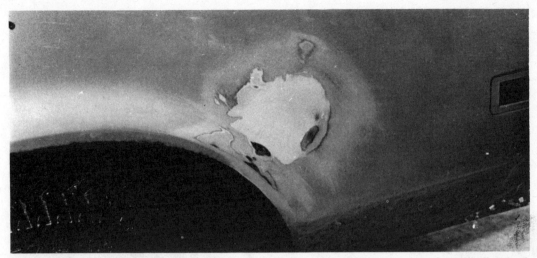

Step 10. When you're finished, the repair should look like this. Note the sand marks extending 2—3 inches out from the repaired area

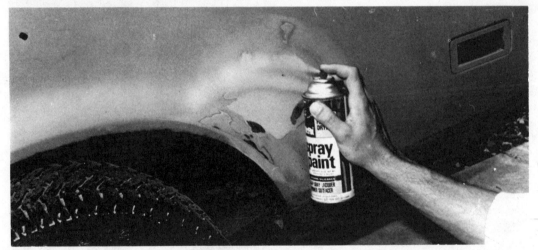

Step 11. Prime the entire area with automotive primer

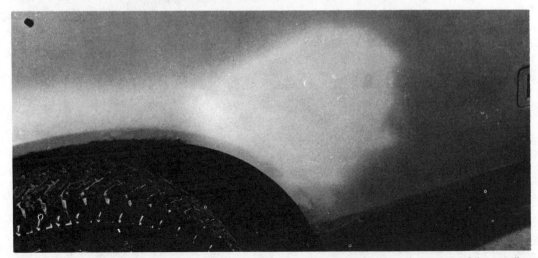

Step 12. The finished repair ready for the final paint coat. Note that the primer has covered the sanding marks (see Step 10). A repair of this size should be able to be spotpainted with good results

REPAIRING RUST HOLES

One thing you have to remember about rust: even if you grind away all the rusted metal in a panel, and repair the area with any of the kits available, *eventually* the rust will return. There are two reasons for this. One, rust is a chemical reaction that causes pressure under the repair from the inside out. That's how the blisters form. Two, the back side of the panel (and the repair) is wide open to moisture, and unpainted body filler acts like a sponge. That's why the best solution to rust problems is to remove the rusted panel and install a new one or have the rusted area cut out and a new piece of sheet metal welded in its place. The trouble with welding is the expense; sometimes it will cost more than the car or truck is worth.

One of the better solutions to do-it-yourself rust repair is the process using a fiberglass cloth repair kit (shown here). This will give a strong repair that resists cracking and moisture and is relatively easy to use. It can be used on large or small holes and also can be applied over contoured surfaces.

Step 1. Rust areas such as this are common and are easily fixed

Step 2. Grind away all traces of rust with a 24-grit grinding disc. Be sure to grind back 3—4 inches from the edge of the hole down to bare metal and be sure all traces of rust are removed

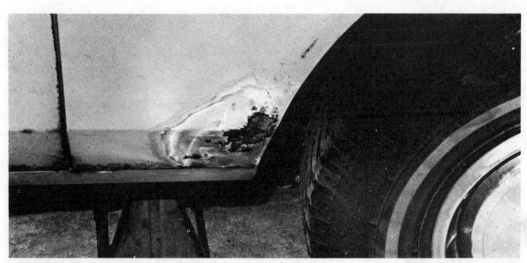

Step 3. Be sure all rust is removed from the edges of the metal. The edges must be ground back to un-rusted metal

Step 4. If you are going to use release film, cut a piece about 2″ larger than the area you have sanded. Place the film over the repair and mark the sanded area on the film. Avoid any unnecessary wrinkling of the film

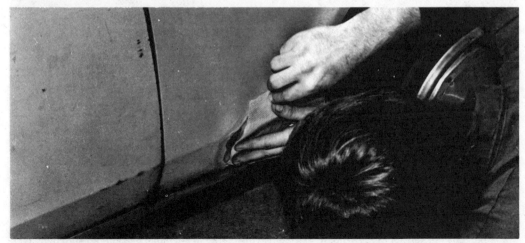

Step 5. Cut 2 pieces of fiberglass matte. One piece should be about 1″ smaller than the sanded area and the second piece should be 1″ smaller than the first. Use sharp scissors to avoid loose ends

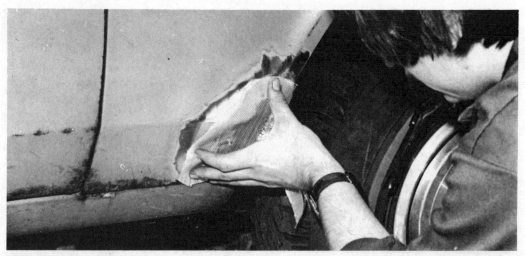

Step 6. Check the dimensions of the release film and cloth by holding them up to the repair area

Step 7. Mix enough repair jelly and cream hardener in the mixing tray to saturate the fiberglass material or fill the repair area. Follow the directions on the container

Step 8. Lay the release sheet on a flat surface and spread an even layer of filler, large enough to cover the repair. Lay the smaller piece of fiberglass cloth in the center of the sheet and spread another layer of repair jelly over the fiberglass cloth. Repeat the operation for the larger piece of cloth. If the fiberglass cloth is not used, spread the repair jelly on the release film, concentrated in the middle of the repair

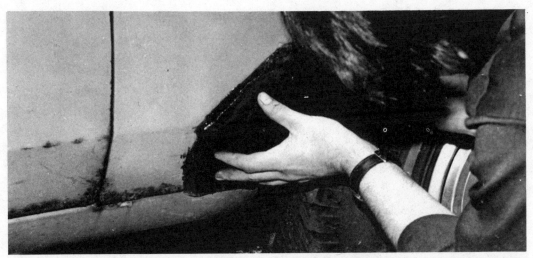

Step 9. Place the repair material over the repair area, with the release film facing outward

Step 10. Use a spreader and work from the center outward to smooth the material, following the body contours. Be sure to remove all air bubbles

Step 11. Wait until the repair has dried tack-free and peel off the release sheet. The ideal working temperature is 65—90° F. Cooler or warmer temperatures or high humidity may require additional curing time

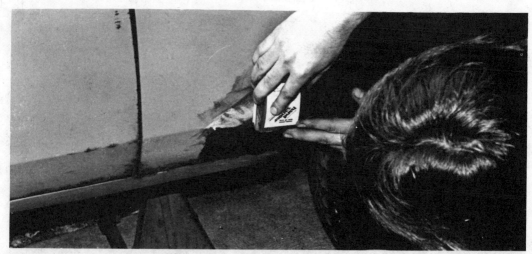

Step 12. Sand and feather-edge the entire area. The initial sanding can be done with a sanding disc on an electric drill if care is used. Finish the sanding with a block sander

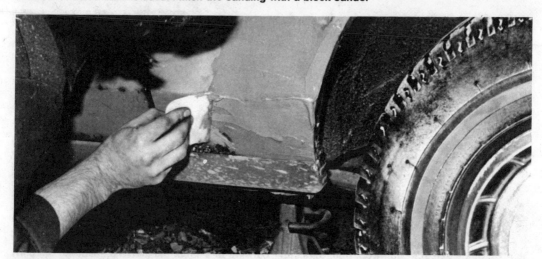

Step 13. When the area is sanded smooth, mix some topcoat and hardener and apply it directly with a spreader. This will give a smooth finish and prevent the glass matte from showing through the paint

Step 14. Block sand the topcoat with finishing sandpaper

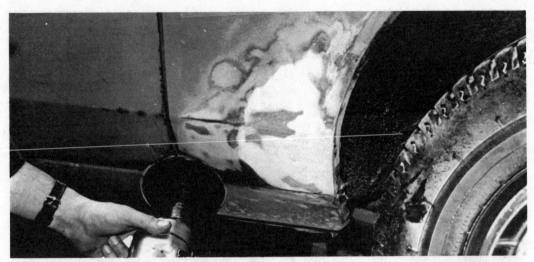

Step 15. To finish this repair, grind out the surface rust along the top edge of the rocker panel

Step 16. Mix some more repair jelly and cream hardener and apply it directly over the surface

Step 17. When it dries tack-free, block sand the surface smooth

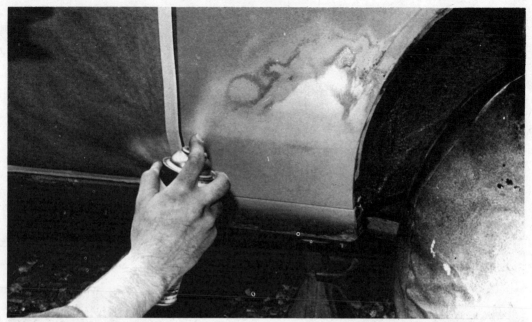

Step 18. If necessary, mask off adjacent panels and spray the entire repair with primer. You are now ready for a color coat

AUTO BODY CARE

There are hundreds—maybe thousands—of products on the market, all designed to protect or aid your car's finish in some manner. There are as many different products as there are ways to use them, but they all have one thing in common—the surface must be clean.

Washing

The primary ingredient for washing your car is water, preferably "soft" water. In many areas of the country, the local water supply is "hard" containing many minerals. The little rings or film that is left on your car's surface after it has dried is the result of "hard" water.

Since you usually can't change the local water supply, the next best thing is to dry the surface before it has a chance to dry itself.

Into the water you usually add soap. Don't use detergents or common, coarse soaps. Your car's paint never truly dries out, but is always evaporating residual oils into the air. Harsh detergents will remove these oils, causing the paint to dry faster than normal. Instead use warm water and a non-detergent soap made especially for waxed surfaces or a liquid soap made for waxed surfaces or a liquid soap made for washing dishes by hand.

Other products that can be used on painted surfaces include baking soda or plain soda water for stubborn dirt.

Wash the car completely, starting at the top, and rinse it completely clean. Abrasive grit should be loaded off under water pressure; scrubbing grit off will scratch the finish. The best washing tool is a sponge, cleaning mitt or soft towel. Whichever you choose, replace it often as each tends to absorb grease and dirt.

Other ways to get a better wash include:

• Don't wash your car in the sun or when the finish is hot.

• Use water pressure to remove caked-on dirt.

• Remove tree-sap and bird effluence immediately. Such substances will eat through wax, polish and paint.

One of the best implements to dry your car is a turkish towel or an old, soft bath towel. Anything with a deep nap will hold any dirt in suspension and not grind it into the paint.

Harder cloths will only grind the grit into the paint making more scratches. Always start drying at the top, followed by the hood and trunk and sides. You'll find there's always more dirt near the rocker panels and wheelwells which will wind up on the rest of the car if you dry these areas first.

Cleaners, Waxes and Polishes

Before going any farther you should know the function of various products.

Cleaners—remove the top layer of dead pigment or paint.

Rubbing or polishing compounds—used to remove stubborn dirt, get rid of minor scratches, smooth away imperfections and partially restore badly weathered paint.

Polishes—contain no abrasives or waxes; they shine the paint by adding oils to the paint.

Waxes—are a protective coating for the polish.

CLEANERS AND COMPOUNDS

Before you apply any wax, you'll have to remove oxidation, road film and other types of pollutants that washing alone will not remove.

The paint on your car never dries completely. There are always residual oils evaporating from the paint into the air. When enough oils are present in the paint, it has a healthy shine (gloss). When too many oils evaporate the paint takes on a whitish cast known as oxidation. The idea of polishing and waxing is to keep enough oil present in the painted surface to prevent oxidation; but when it occurs, the only recourse is to remove the top layer of "dead" paint, exposing the healthy paint underneath.

Products to remove oxidation and road film are sold under a variety of generic names—polishes, cleaner, rubbing compound, cleaner/polish, polish/cleaner, self-polishing wax, pre-wax cleaner, finish restorer and many more. Regardless of name there are two types of cleaners—abrasive cleaners (sometimes called polishing or rubbing compounds) that remove oxidation by grinding away the top layer of "dead" paint, or chemical cleaners that dissolve the "dead" pigment, allowing it to be wiped away.

Abrasive cleaners, by their nature, leave thousands of minute scratches in the finish, which must be polished out later. These should only be used in extreme cases, but are usually the only thing to use on badly oxidized paint finishes. Chemical cleaners are much milder but are not strong enough for severe cases of oxidation or weathered paint.

The most popular cleaners are liquid or paste abrasive polishing and rubbing compounds. Polishing compounds have a finer abrasive grit for medium duty work. Rubbing compounds are a coarser abrasive and for heavy duty work. Unless you are familiar with how to use compounds, be very careful. Excessive rubbing with any type of compound or cleaner can grind right through the paint to primer or bare metal. Follow the directions on the container—depending on type, the cleaner may or may not be OK for your paint. For example, some cleaners are not formulated for acrylic lacquer finishes.

When a small area needs compounding or heavy polishing, it's best to do the job by hand. Some people prefer a powered buffer for large areas. Avoid cutting through the paint along styling edges on the body. Small, hand operations where the compound is applied and rubbed using cloth folded into a thick ball allow you to work in straight lines along such edges.

To avoid cutting through on the edges when using a power buffer, try masking tape. Just cover the edge with tape while using power. Then finish the job by hand with the tape removed. Even then work carefully. The paint tends to be a lot thinner along the sharp ridges stamped into the panels.

Whether compounding by machine or by hand, only work on a small area and apply the compound sparingly. If the materials are spread too thin, or allowed to sit too long, they dry out. Once dry they lose the ability to deliver a smooth, clean finish. Also, dried out polish tends to cause the buffer to stick in one spot. This in turn can burn or cut through the finish.

WAXES AND POLISHES

Your car's finish can be protected in a number of ways. A cleaner/wax or polish/cleaner followed by wax or variations of each all provide good results. The two-step approach (polish followed by wax) is probably slightly better but consumes more time and effort. Properly fed with oils, your paint should never need cleaning, but despite the best polishing job, it won't last unless it's protected with wax. Without wax, polish must be renewed at least once a month to prevent oxidation. Years ago (some still swear by it today), the best wax was made from the Brazilian palm, the Carnuba, favored for its vegetable base and high melting point. However, modern synthetic waxes are harder, which means they protect against moisture better, and chemically inert silicone is used for a long lasting protection. The only problem with silicone wax is that it penetrates all layers of paint. To repaint or touch up a panel or car protected by silicone wax, you have to completely strip the finish to avoid "fish-eyes."

Under normal conditions, silicone waxes will last 4–6 months, but you have to be careful of wax build-up from too much waxing. Too thick a coat of wax is just as bad as no wax at all; it stops the paint from breathing.

Combination cleaners/waxes have become popular lately because they remove the old layer of wax plus light oxidation, while putting on a fresh coat of wax at the same time. Some cleaners/waxes contain abrasive cleaners which require caution, although many cleaner/waxes use a chemical cleaner.

Applying Wax or Polish

You may view polishing and waxing your car as a pleasant way to spend an afternoon, or as a boring chore, but it has to be done to keep the paint on your car. Caring for the paint doesn't require special tools, but you should follow a few rules.

1. Use a good quality wax.

2. Before applying any wax or polish, be sure the surface is completely clean. Just because the car looks clean, doesn't mean it's ready for polish or wax.

3. If the finish on your car is weathered, dull, or oxidized, it will probably have to be compounded to remove the old or oxidized paint. If the paint is simply dulled from lack of care, one of the non-abrasive cleaners known as polishing compounds will do the trick. If the paint is severely scratched or really dull, you'll probably have to use a rubbing compound to prepare the finish for waxing. If you're not sure which one to use, use the polishing compound, since you can easily ruin the finish by using too strong a compound.

4. Don't apply wax, polish or compound in direct sunlight, even if the directions on the can say you can. Most waxes will not cure properly in bright sunlight and you'll probably end up with a blotchy looking finish.

5. Don't rub the wax off too soon. The result will be a wet, dull looking finish. Let the wax dry thoroughly before buffing it off.

6. A constant debate among car enthusiasts is how wax should be applied. Some maintain pastes or liquids should be applied in a circular motion, but body shop experts have long thought that this approach results in barely detectable circular abrasions, especially on cars that are waxed frequently. They advise rubbing in straight lines, especially if any kind of cleaner is involved.

7. If an applicator is not supplied with the wax, use a piece of soft cheesecloth or very soft lint-free material. The same applies to buffing the surface.

SPECIAL SURFACES

One-step combination cleaner and wax formulas shouldn't be used on many of the special surfaces which abound on cars. The one-step materials contain abrasives to achieve a clean surface under the wax top coat. The abrasives are so mild that you could clean a car every week for a couple of years without fear of rubbing through the paint. But this same level of abrasiveness might, through repeated use, damage decals used for special trim effects. This includes wide stripes, wood-grain trim and other appliques.

Painted plastics must be cleaned with care. If a cleaner is too aggressive it will cut through the paint and expose the primer. If bright trim such as polished aluminum or chrome is painted, cleaning must be performed with even greater care. If rubbing compound is being used, it will cut faster than polish.

Abrasive cleaners will dull an acrylic finish. The best way to clean these newer finishes is with a non-abrasive liquid polish. Only dirt and oxidation, not paint, will be removed.

Taking a few minutes to read the instructions on the can of polish or wax will help prevent making serious mistakes. Not all preparations will work on all surfaces. And some are intended for power application while others will only work when applied by hand.

Don't get the idea that just pouring on some polish and then hitting it with a buffer will suffice. Power equipment speeds the operation. But it also adds a measure of risk. It's very easy to damage the finish if you use the wrong methods or materials.

Caring for Chrome

Read the label on the container. Many products are formulated specifically for chrome, but others contain abrasives that will scratch the chrome finish. If it isn't recommended for chrome, don't use it.

Never use steel wool or kitchen soap pads to clean chrome. Be careful not to get chrome cleaner on paint or interior vinyl surfaces. If you do, get it off immediately.

Troubleshooting

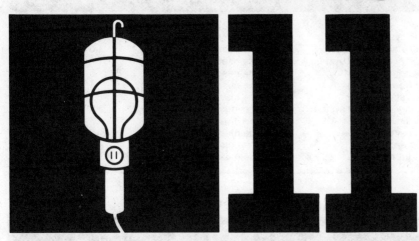

This section is designed to aid in the quick, accurate diagnosis of automotive problems. While automotive repairs can be made by many people, accurate troubleshooting is a rare skill for the amateur and professional alike.

In its simplest state, troubleshooting is an exercise in logic. It is essential to realize that an automobile is really composed of a series of systems. Some of these systems are interrelated; others are not. Automobiles operate within a framework of logical rules and physical laws, and the key to troubleshooting is a good understanding of all the automotive systems.

This section breaks the car or truck down into its component systems, allowing the problem to be isolated. The charts and diagnostic road maps list the most common problems and the most probable causes of trouble. Obviously it would be impossible to list every possible problem that could happen along with every possible cause, but it will locate MOST problems and eliminate a lot of unnecessary guesswork. The systematic format will locate problems within a given system, but, because many automotive systems are interrelated, the solution to your particular problem may be found in a number of systems on the car or truck.

USING THE TROUBLESHOOTING CHARTS

This book contains all of the specific information that the average do-it-yourself mechanic needs to repair and maintain his or her car or truck. The troubleshooting charts are designed to be used in conjunction with the specific procedures and information in the text. For instance, troubleshooting a point-type ignition system is fairly standard for all models, but you may be directed to the text to find procedures for troubleshooting an individual type of electronic ignition. You will also have to refer to the specification charts throughout the book for specifications applicable to your car or truck.

TOOLS AND EQUIPMENT

The tools illustrated in Chapter 1 (plus two more diagnostic pieces) will be adequate to troubleshoot most problems. The two other tools needed are a voltmeter and an ohmmeter. These can be purchased separately or in combination, known as a VOM meter.

In the event that other tools are required, they will be noted in the procedures.

Troubleshooting Engine Problems

See Chapters 2, 3, 4 for more information and service procedures.

Index to Systems

System	To Test	Group
Battery	Engine need not be running	1
Starting system	Engine need not be running	2
Primary electrical system	Engine need not be running	3
Secondary electrical system	Engine need not be running	4
Fuel system	Engine need not be running	5
Engine compression	Engine need not be running	6
Engine vacuum	Engine must be running	7
Secondary electrical system	Engine must be running	8
Valve train	Engine must be running	9
Exhaust system	Engine must be running	10
Cooling system	Engine must be running	11
Engine lubrication	Engine must be running	12

Index to Problems

Problem: Symptom	Begin at Specific Diagnosis, Number
Engine Won't Start:	
Starter doesn't turn	1.1, 2.1
Starter turns, engine doesn't	2.1
Starter turns engine very slowly	1.1, 2.4
Starter turns engine normally	3.1, 4.1
Starter turns engine very quickly	6.1
Engine fires intermittently	4.1
Engine fires consistently	5.1, 6.1
Engine Runs Poorly:	
Hard starting	3.1, 4.1, 5.1, 8.1
Rough idle	4.1, 5.1, 8.1
Stalling	3.1, 4.1, 5.1, 8.1
Engine dies at high speeds	4.1, 5.1
Hesitation (on acceleration from standing stop)	5.1, 8.1
Poor pickup	4.1, 5.1, 8.1
Lack of power	3.1, 4.1, 5.1, 8.1
Backfire through the carburetor	4.1, 8.1, 9.1
Backfire through the exhaust	4.1, 8.1, 9.1
Blue exhaust gases	6.1, 7.1
Black exhaust gases	5.1
Running on (after the ignition is shut off)	3.1, 8.1
Susceptible to moisture	4.1
Engine misfires under load	4.1, 7.1, 8.4, 9.1
Engine misfires at speed	4.1, 8.4
Engine misfires at idle	3.1, 4.1, 5.1, 7.1, 8.4

Sample Section

Test and Procedure	Results and Indications	Proceed to
4.1—Check for spark: Hold each spark plug wire approximately ¼″ from ground with gloves or a heavy, dry rag. Crank the engine and observe the spark.	If no spark is evident:	→**4.2**
	If spark is good in some cases:	→**4.3**
	If spark is good in all cases:	→**4.6**

Specific Diagnosis

This section is arranged so that following each test, instructions are given to proceed to another, until a problem is diagnosed.

Section 1—Battery

Test and Procedure	Results and Indications	Proceed to
1.1—Inspect the battery visually for case condition (corrosion, cracks) and water level.	If case is cracked, replace battery:	**1.4**
	If the case is intact, remove corrosion with a solution of baking soda and water (**CAUTION:** *do not get the solution into the battery*), and fill with water:	**1.2**

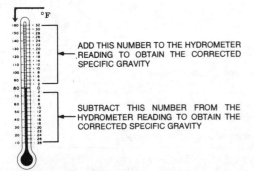

DIRT ON TOP OF BATTERY
CORROSION
PLUGGED VENT
LOOSE CABLE OR POSTS
CRACKS
LOW WATER LEVEL

Inspect the battery case

Test and Procedure	Results and Indications	Proceed to
1.2—Check the battery cable connections: Insert a screwdriver between the battery post and the cable clamp. Turn the headlights on high beam, and observe them as the screwdriver is gently twisted to ensure good metal to metal contact.	If the lights brighten, remove and clean the clamp and post; coat the post with petroleum jelly, install and tighten the clamp:	**1.4**
	If no improvement is noted:	**1.3**

TESTING BATTERY CABLE CONNECTIONS USING A SCREWDRIVER

Test and Procedure	Results and Indications	Proceed to
1.3—Test the state of charge of the battery using an individual cell tester or hydrometer.	If indicated, charge the battery. **NOTE:** *If no obvious reason exists for the low state of charge (i.e., battery age, prolonged storage), proceed to:*	**1.4**

°F

ADD THIS NUMBER TO THE HYDROMETER READING TO OBTAIN THE CORRECTED SPECIFIC GRAVITY

SUBTRACT THIS NUMBER FROM THE HYDROMETER READING TO OBTAIN THE CORRECTED SPECIFIC GRAVITY

Specific Gravity (@ 80° F.)

Minimum	Battery Charge
1.260	100% Charged
1.230	75% Charged
1.200	50% Charged
1.170	25% Charged
1.140	Very Little Power Left
1.110	Completely Discharged

The effects of temperature on battery specific gravity (left) and amount of battery charge in relation to specific gravity (right)

Test and Procedure	Results and Indications	Proceed to
1.4—Visually inspect battery cables for cracking, bad connection to ground, or bad connection to starter.	If necessary, tighten connections or replace the cables:	**2.1**

Section 2—Starting System
See Chapter 3 for service procedures

Test and Procedure	Results and Indications	Proceed to
Note: Tests in Group 2 are performed with coil high tension lead disconnected to prevent accidental starting.		
2.1—Test the starter motor and solenoid: Connect a jumper from the battery post of the solenoid (or relay) to the starter post of the solenoid (or relay).	If starter turns the engine normally:	**2.2**
	If the starter buzzes, or turns the engine very slowly:	**2.4**
	If no response, replace the solenoid (or relay).	**3.1**
	If the starter turns, but the engine doesn't, ensure that the flywheel ring gear is intact. If the gear is undamaged, replace the starter drive.	**3.1**
2.2—Determine whether ignition override switches are functioning properly (clutch start switch, neutral safety switch), by connecting a jumper across the switch(es), and turning the ignition switch to "start".	If starter operates, adjust or replace switch:	**3.1**
	If the starter doesn't operate:	**2.3**
2.3—Check the ignition switch "start" position: Connect a 12V test lamp or voltmeter between the starter post of the solenoid (or relay) and ground. Turn the ignition switch to the "start" position, and jiggle the key.	If the lamp doesn't light or the meter needle doesn't move when the switch is turned, check the ignition switch for loose connections, cracked insulation, or broken wires. Repair or replace as necessary:	**3.1**
	If the lamp flickers or needle moves when the key is jiggled, replace the ignition switch.	**3.3**

Checking the ignition switch "start" position

STARTER RELAY
(IF EQUIPPED)

Test and Procedure	Results and Indications	Proceed to
2.4—Remove and bench test the starter, according to specifications in the engine electrical section.	If the starter does not meet specifications, repair or replace as needed:	**3.1**
	If the starter is operating properly:	**2.5**
2.5—Determine whether the engine can turn freely: Remove the spark plugs, and check for water in the cylinders. Check for water on the dipstick, or oil in the radiator. Attempt to turn the engine using an 18″ flex drive and socket on the crankshaft pulley nut or bolt.	If the engine will turn freely only with the spark plugs out, and hydrostatic lock (water in the cylinders) is ruled out, check valve timing:	**9.2**
	If engine will not turn freely, and it is known that the clutch and transmission are free, the engine must be disassembled for further evaluation:	**Chapter 3**

Section 3—Primary Electrical System

Test and Procedure	Results and Indications	Proceed to
3.1—Check the ignition switch "on" position: Connect a jumper wire between the distributor side of the coil and ground, and a 12V test lamp between the switch side of the coil and ground. Remove the high tension lead from the coil. Turn the ignition switch on and jiggle the key.	If the lamp lights:	**3.2**
	If the lamp flickers when the key is jiggled, replace the ignition switch:	**3.3**
	If the lamp doesn't light, check for loose or open connections. If none are found, remove the ignition switch and check for continuity. If the switch is faulty, replace it:	**3.3**

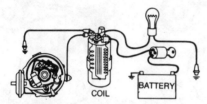

Checking the ignition switch "on" position

3.2—Check the ballast resistor or resistance wire for an open circuit, using an ohmmeter. See Chapter 3 for specific tests.	Replace the resistor or resistance wire if the resistance is zero. **NOTE:** *Some ignition systems have no ballast resistor.*	**3.3**

 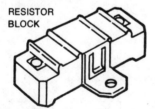

RESISTOR BLOCK

CALIBRATED RESISTANCE LEAD

Two types of resistors

3.3—On point-type ignition systems, visually inspect the breaker points for burning, pitting or excessive wear. Gray coloring of the point contact surfaces is normal. Rotate the crankshaft until the contact heel rests on a high point of the distributor cam and adjust the point gap to specifications. On electronic ignition models, remove the distributor cap and visually inspect the armature. Ensure that the armature pin is in place, and that the armature is on tight and rotates when the engine is cranked. Make sure there are no cracks, chips or rounded edges on the armature.	If the breaker points are intact, clean the contact surfaces with fine emery cloth, and adjust the point gap to specifications. If the points are worn, replace them. On electronic systems, replace any parts which appear defective. If condition persists:	**3.4**

Test and Procedure	Results and Indications	Proceed to
3.4—On point-type ignition systems, connect a dwell-meter between the distributor primary lead and ground. Crank the engine and observe the point dwell angle. On electronic ignition systems, conduct a stator (magnetic pickup assembly) test. See Chapter 3.	On point-type systems, adjust the dwell angle if necessary. **NOTE:** *Increasing the point gap decreases the dwell angle and vice-versa.*	**3.6**
	If the dwell meter shows little or no reading;	**3.5**
	On electronic ignition systems, if the stator is bad, replace the stator. If the stator is good, proceed to the other tests in Chapter 3.	

WIDE GAP   NARROW GAP

CLOSE   OPEN

SMALL DWELL   LARGE DWELL

NORMAL DWELL

INSUFFICIENT DWELL

EXCESSIVE DWELL

Dwell is a function of point gap

3.5—On the point-type ignition systems, check the condenser for short: connect an ohmmeter across the condenser body and the pigtail lead.	If any reading other than infinite is noted, replace the condenser	**3.6**

OHMMETER

Checking the condenser for short

3.6—Test the coil primary resistance: On point-type ignition systems, connect an ohmmeter across the coil primary terminals, and read the resistance on the low scale. Note whether an external ballast resistor or resistance wire is used. On electronic ignition systems, test the coil primary resistance as in Chapter 3.	Point-type ignition coils utilizing ballast resistors or resistance wires should have approximately 1.0 ohms resistance. Coils with internal resistors should have approximately 4.0 ohms resistance. If values far from the above are noted, replace the coil.	**4.1**

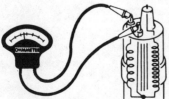

Check the coil primary resistance

Section 4—Secondary Electrical System

See Chapters 2–3 for service procedures

Test and Procedure	Results and Indications	Proceed to
4.1—Check for spark: Hold each spark plug wire approximately ¼″ from ground with gloves or a heavy, dry rag. Crank the engine, and observe the spark.	If no spark is evident:	**4.2**
	If spark is good in some cylinders:	**4.3**
	If spark is good in all cylinders:	**4.6**

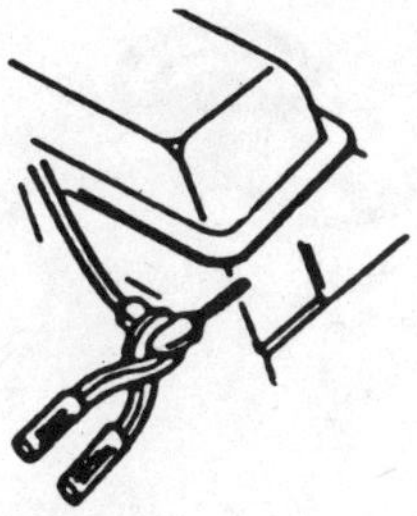

Check for spark at the plugs

Test and Procedure	Results and Indications	Proceed to
4.2—Check for spark at the coil high tension lead: Remove the coil high tension lead from the distributor and position it approximately ¼″ from ground. Crank the engine and observe spark. **CAUTION:** *This test should not be performed on engines equipped with electronic ignition.*	If the spark is good and consistent:	**4.3**
	If the spark is good but intermittent, test the primary electrical system starting at 3.3:	**3.3**
	If the spark is weak or non-existent, replace the coil high tension lead, clean and tighten all connections and retest. If no improvement is noted:	**4.4**
4.3—Visually inspect the distributor cap and rotor for burned or corroded contacts, cracks, carbon tracks, or moisture. Also check the fit of the rotor on the distributor shaft (where applicable).	If moisture is present, dry thoroughly, and retest per 4.1:	**4.1**
	If burned or excessively corroded contacts, cracks, or carbon tracks are noted, replace the defective part(s) and retest per 4.1:	**4.1**
	If the rotor and cap appear intact, or are only slightly corroded, clean the contacts thoroughly (including the cap towers and spark plug wire ends) and retest per 4.1: If the spark is good in all cases:	**4.6**
	If the spark is poor in all cases:	**4.5**

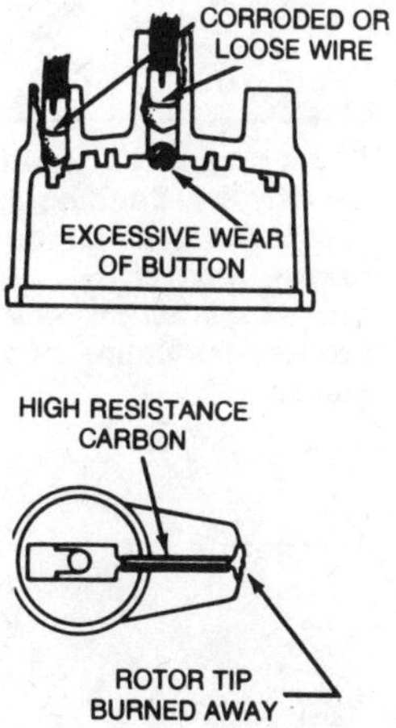

CORRODED OR LOOSE WIRE

EXCESSIVE WEAR OF BUTTON

HIGH RESISTANCE CARBON

ROTOR TIP BURNED AWAY

Inspect the distributor cap and rotor

Test and Procedure	Results and Indications	Proceed to
4.4—Check the coil secondary resistance: On point-type systems connect an ohmmeter across the distributor side of the coil and the coil tower. Read the resistance on the high scale of the ohmmeter. On electronic ignition systems, see Chapter 3 for specific tests.	The resistance of a satisfactory coil should be between 4,000 and 10,000 ohms. If resistance is considerably higher (i.e., 40,000 ohms) replace the coil and retest per 4.1. **NOTE:** *This does not apply to high performance coils.*	

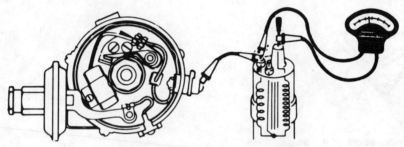

Testing the coil secondary resistance

| **4.5**—Visually inspect the spark plug wires for cracking or brittleness. Ensure that no two wires are positioned so as to cause induction firing (adjacent and parallel). Remove each wire, one by one, and check resistance with an ohmmeter. | Replace any cracked or brittle wires. If any of the wires are defective, replace the entire set. Replace any wires with excessive resistance (over 8000 Ω per foot for suppression wire), and separate any wires that might cause induction firing. | **4.6** |

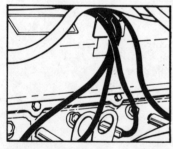

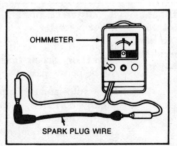

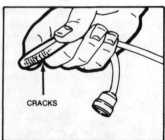

Misfiring can be the result of spark plug leads to adjacent, consecutively firing cylinders running parallel and too close together

On point-type ignition systems, check the spark plug wires as shown. On electronic ignitions, do not remove the wire from the distributor cap terminal; instead, test through the cap

Spark plug wires can be checked visually by bending them in a loop over your finger. This will reveal any cracks, burned or broken insulation. Any wire with cracked insulation should be replaced

| **4.6**—Remove the spark plugs, noting the cylinders from which they were removed, and evaluate according to the color photos in the middle of this book. | See following. | **See following.** |

Test and Procedure	Results and Indications	Proceed to
4.7—Examine the location of all the plugs.	The following diagrams illustrate some of the conditions that the location of plugs will reveal.	4.8

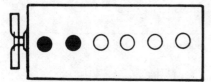

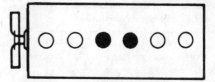

Two adjacent plugs are fouled in a 6-cylinder engine, 4-cylinder engine or either bank of a V-8. This is probably due to a blown head gasket between the two cylinders

The two center plugs in a 6-cylinder engine are fouled. Raw fuel may be "boiled" out of the carburetor into the intake manifold after the engine is shut-off. Stop-start driving can also foul the center plugs, due to overly rich mixture. Proper float level, a new float needle and seat or use of an insulating spacer may help this problem

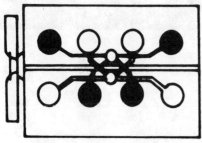

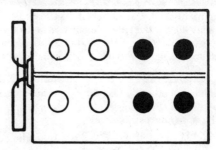

An unbalanced carburetor is indicated. Following the fuel flow on this particular design shows that the cylinders fed by the right-hand barrel are fouled from overly rich mixture, while the cylinders fed by the left-hand barrel are normal

If the four rear plugs are overheated, a cooling system problem is suggested. A thorough cleaning of the cooling system may restore coolant circulation and cure the problem

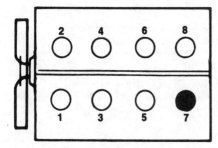

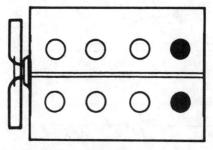

Finding one plug overheated may indicate an intake manifold leak near the affected cylinder. If the overheated plug is the second of two adjacent, consecutively firing plugs, it could be the result of ignition cross-firing. Separating the leads to these two plugs will eliminate cross-fire

Occasionally, the two rear plugs in large, lightly used V-8's will become oil fouled. High oil consumption and smoky exhaust may also be noticed. It is probably due to plugged oil drain holes in the rear of the cylinder head, causing oil to be sucked in around the valve stems. This usually occurs in the rear cylinders first, because the engine slants that way

Test and Procedure	Results and Indications	Proceed to
4.8—Determine the static ignition timing. Using the crankshaft pulley timing marks as a guide, locate top dead center on the compression stroke of the number one cylinder.	The rotor should be pointing toward the No. 1 tower in the distributor cap, and, on electronic ignitions, the armature spoke for that cylinder should be lined up with the stator.	4.8
4.9—Check coil polarity: Connect a voltmeter negative lead to the coil high tension lead, and the positive lead to ground (**NOTE:** *Reverse the hook-up for positive ground systems*). Crank the engine momentarily.	If the voltmeter reads up-scale, the polarity is correct:	5.1
	If the voltmeter reads down-scale, reverse the coil polarity (switch the primary leads):	5.1

Checking coil polarity

Section 5—Fuel System
See Chapter 4 for service procedures

Test and Procedure	Results and Indications	Proceed to
5.1—Determine that the air filter is functioning efficiently: Hold paper elements up to a strong light, and attempt to see light through the filter.	Clean permanent air filters in solvent (or manufacturer's recommendation), and allow to dry. Replace paper elements through which light cannot be seen:	5.2
5.2—Determine whether a flooding condition exists: Flooding is identified by a strong gasoline odor, and excessive gasoline present in the throttle bore(s) of the carburetor.	If flooding is not evident:	5.3
	If flooding is evident, permit the gasoline to dry for a few moments and restart. If flooding doesn't recur:	5.7
	If flooding is persistent:	5.5

If the engine floods repeatedly, check the choke butterfly flap

5.3—Check that fuel is reaching the carburetor: Detach the fuel line at the carburetor inlet. Hold the end of the line in a cup (not styrofoam), and crank the engine.	If fuel flows smoothly:	5.7
	If fuel doesn't flow (**NOTE:** *Make sure that there is fuel in the tank*), or flows erratically:	5.4

Check the fuel pump by disconnecting the output line (fuel pump-to-carburetor) at the carburetor and operating the starter briefly

Test and Procedure	Results and Indications	Proceed to
5.4—Test the fuel pump: Disconnect all fuel lines from the fuel pump. Hold a finger over the input fitting, crank the engine (with electric pump, turn the ignition or pump on); and feel for suction.	If suction is evident, blow out the fuel line to the tank with low pressure compressed air until bubbling is heard from the fuel filler neck. Also blow out the carburetor fuel line (both ends disconnected):	**5.7**
	If no suction is evident, replace or repair the fuel pump: **NOTE:** *Repeated oil fouling of the spark plugs, or a no-start condition, could be the result of a ruptured vacuum booster pump diaphragm, through which oil or gasoline is being drawn into the intake manifold (where applicable).*	**5.7**
5.5—Occasionally, small specks of dirt will clog the small jets and orifices in the carburetor. With the engine cold, hold a flat piece of wood or similar material over the carburetor, where possible, and crank the engine.	If the engine starts, but runs roughly the engine is probably not run enough. If the engine won't start:	**5.9**
5.6—Check the needle and seat: Tap the carburetor in the area of the needle and seat.	If flooding stops, a gasoline additive (e.g., Gumout) will often cure the problem:	**5.7**
	If flooding continues, check the fuel pump for excessive pressure at the carburetor (according to specifications). If the pressure is normal, the needle and seat must be removed and checked, and/or the float level adjusted:	**5.7**
5.7—Test the accelerator pump by looking into the throttle bores while operating the throttle.	If the accelerator pump appears to be operating normally:	**5.8**
	If the accelerator pump is not operating, the pump must be reconditioned. Where possible, service the pump with the carburetor(s) installed on the engine. If necessary, remove the carburetor. Prior to removal:	**5.8**

Check for gas at the carburetor by looking down the carburetor throat while someone moves the accelerator

Test and Procedure	Results and Indications	Proceed to
5.8—Determine whether the carburetor main fuel system is functioning: Spray a commercial starting fluid into the carburetor while attempting to start the engine.	If the engine starts, runs for a few seconds, and dies:	**5.9**
	If the engine doesn't start:	**6.1**

Test and Procedure	Results and Indications	Proceed to
5.9—Uncommon fuel system malfunctions: See below:	If the problem is solved:	6.1
	If the problem remains, remove and recondition the carburetor.	

Condition	Indication	Test	Prevailing Weather Conditions	Remedy
Vapor lock	Engine will not restart shortly after running.	Cool the components of the fuel system until the engine starts. Vapor lock can be cured faster by draping a wet cloth over a mechanical fuel pump.	Hot to very hot	Ensure that the exhaust manifold heat control valve is operating. Check with the vehicle manufacturer for the recommended solution to vapor lock on the model in question.
Carburetor icing	Engine will not idle, stalls at low speeds.	Visually inspect the throttle plate area of the throttle bores for frost.	High humidity, 32–40° F.	Ensure that the exhaust manifold heat control valve is operating, and that the intake manifold heat riser is not blocked.
Water in the fuel	Engine sputters and stalls; may not start.	Pump a small amount of fuel into a glass jar. Allow to stand, and inspect for droplets or a layer of water.	High humidity, extreme temperature changes.	For droplets, use one or two cans of commercial gas line anti-freeze. For a layer of water, the tank must be drained, and the fuel lines blown out with compressed air.

Section 6—Engine Compression
See Chapter 3 for service procedures

6.1—Test engine compression: Remove all spark plugs. Block the throttle wide open. Insert a compression gauge into a spark plug port, crank the engine to obtain the maximum reading, and record.	If compression is within limits on all cylinders:	7.1
	If gauge reading is extremely low on all cylinders:	6.2
	If gauge reading is low on one or two cylinders: (If gauge readings are identical and low on two or more adjacent cylinders, the head gasket must be replaced.)	6.2

Checking compression

6.2—Test engine compression (wet): Squirt approximately 30 cc. of engine oil into each cylinder, and retest per 6.1.	If the readings improve, worn or cracked rings or broken pistons are indicated:	See Chapter 3
	If the readings do not improve, burned or excessively carboned valves or a jumped timing chain are indicated:	7.1
	NOTE: A jumped timing chain is often indicated by difficult cranking.	

Section 7—Engine Vacuum
See Chapter 3 for service procedures

Test and Procedure	Results and Indications	Proceed to
7.1—Attach a vacuum gauge to the intake manifold beyond the throttle plate. Start the engine, and observe the action of the needle over the range of engine speeds.	See below.	**See below**

INDICATION: normal engine in good condition

Proceed to: 8.1

Normal engine
Gauge reading: steady, from 17–22 in./Hg.

INDICATION: sticking valves or ignition miss

Proceed to: 9.1, 8.3

Sticking valves
Gauge reading: intermittent fluctuation at idle

INDICATION: late ignition or valve timing, low compression, stuck throttle valve, leaking carburetor or manifold gasket

Proceed to: 6.1

Incorrect valve timing
Gauge reading: low (10–15 in./Hg) but steady

INDICATION: improper carburetor adjustment or minor intake leak.

Proceed to: 7.2

Carburetor requires adjustment
Gauge reading: drifting needle

INDICATION: ignition miss, blown cylinder head gasket, leaking valve or weak valve spring

Proceed to: 8.3, 6.1

Blown head gasket
Gauge reading: needle fluctuates as engine speed increases

INDICATION: burnt valve or faulty valve clearance. Needle will fall when defective valve operates

Proceed to: 9.1

Burnt or leaking valves
Gauge reading: steady needle, but drops regularly

INDICATION: choked muffler, excessive back pressure in system

Proceed to: 10.1

Clogged exhaust system
Gauge reading: gradual drop in reading at idle

INDICATION: worn valve guides

Proceed to: 9.1

Worn valve guides
Gauge reading: needle vibrates excessively at idle, but steadies as engine speed increases

White pointer = steady gauge hand   Black pointer = fluctuating gauge hand

Test and Procedure	Results and Indications	Proceed to
7.2—Attach a vacuum gauge per 7.1, and test for an intake manifold leak. Squirt a small amount of oil around the intake manifold gaskets, carburetor gaskets, plugs and fittings. Observe the action of the vacuum gauge.	If the reading improves, replace the indicated gasket, or seal the indicated fitting or plug:	**8.1**
	If the reading remains low:	**7.3**
7.3—Test all vacuum hoses and accessories for leaks as described in 7.2. Also check the carburetor body (dashpots, automatic choke mechanism, throttle shafts) for leaks in the same manner.	If the reading improves, service or replace the offending part(s):	**8.1**
	If the reading remains low:	**6.1**

Section 8—Secondary Electrical System
See Chapter 2 for service procedures

Test and Procedure	Results and Indications	Proceed to
8.1—Remove the distributor cap and check to make sure that the rotor turns when the engine is cranked. Visually inspect the distributor components.	Clean, tighten or replace any components which appear defective.	**8.2**
8.2—Connect a timing light (per manufacturer's recommendation) and check the dynamic ignition timing. Disconnect and plug the vacuum hose(s) to the distributor if specified, start the engine, and observe the timing marks at the specified engine speed.	If the timing is not correct, adjust to specifications by rotating the distributor in the engine: (Advance timing by rotating distributor opposite normal direction of rotor rotation, retard timing by rotating distributor in same direction as rotor rotation.)	**8.3**
8.3—Check the operation of the distributor advance mechanism(s): To test the mechanical advance, disconnect the vacuum lines from the distributor advance unit and observe the timing marks with a timing light as the engine speed is increased from idle. If the mark moves smoothly, without hesitation, it may be assumed that the mechanical advance is functioning properly. To test vacuum advance and/or retard systems, alternately crimp and release the vacuum line, and observe the timing mark for movement. If movement is noted, the system is operating.	If the systems are functioning:	**8.4**
	If the systems are not functioning, remove the distributor, and test on a distributor tester:	**8.4**
8.4—Locate an ignition miss: With the engine running, remove each spark plug wire, one at a time, until one is found that doesn't cause the engine to roughen and slow down.	When the missing cylinder is identified:	**4.1**

Section 9—Valve Train
See Chapter 3 for service procedures

Test and Procedure	Results and Indications	Proceed to
9.1—Evaluate the valve train: Remove the valve cover, and ensure that the valves are adjusted to specifications. A mechanic's stethoscope may be used to aid in the diagnosis of the valve train. By pushing the probe on or near push rods or rockers, valve noise often can be isolated. A timing light also may be used to diagnose valve problems. Connect the light according to manufacturer's recommendations, and start the engine. Vary the firing moment of the light by increasing the engine speed (and therefore the ignition advance), and moving the trigger from cylinder to cylinder. Observe the movement of each valve.	Sticking valves or erratic valve train motion can be observed with the timing light. The cylinder head must be disassembled for repairs.	**See Chapter 3**
9.2—Check the valve timing: Locate top dead center of the No. 1 piston, and install a degree wheel or tape on the crankshaft pulley or damper with zero corresponding to an index mark on the engine. Rotate the crankshaft in its direction of rotation, and observe the opening of the No. 1 cylinder intake valve. The opening should correspond with the correct mark on the degree wheel according to specifications.	If the timing is not correct, the timing cover must be removed for further investigation.	**See Chapter 3**

Section 10—Exhaust System

Test and Procedure	Results and Indications	Proceed to
10.1—Determine whether the exhaust manifold heat control valve is operating: Operate the valve by hand to determine whether it is free to move. If the valve is free, run the engine to operating temperature and observe the action of the valve, to ensure that it is opening.	If the valve sticks, spray it with a suitable solvent, open and close the valve to free it, and retest.	
	If the valve functions properly:	**10.2**
	If the valve does not free, or does not operate, replace the valve:	**10.2**
10.2—Ensure that there are no exhaust restrictions: Visually inspect the exhaust system for kinks, dents, or crushing. Also note that gases are flowing freely from the tailpipe at all engine speeds, indicating no restriction in the muffler or resonator.	Replace any damaged portion of the system:	**11.1**

Section 11—Cooling System

See Chapter 3 for service procedures

Test and Procedure	Results and Indications	Proceed to
11.1—Visually inspect the fan belt for glazing, cracks, and fraying, and replace if necessary. Tighten the belt so that the longest span has approximately ½″ play at its midpoint under thumb pressure (see Chapter 1).	Replace or tighten the fan belt as necessary:	**11.2**

Checking belt tension

Test and Procedure	Results and Indications	Proceed to
11.2—Check the fluid level of the cooling system.	If full or slightly low, fill as necessary:	**11.5**
	If extremely low:	**11.3**
11.3—Visually inspect the external portions of the cooling system (radiator, radiator hoses, thermostat elbow, water pump seals, heater hoses, etc.) for leaks. If none are found, pressurize the cooling system to 14–15 psi.	If cooling system holds the pressure:	**11.5**
	If cooling system loses pressure rapidly, reinspect external parts of the system for leaks under pressure. If none are found, check dipstick for coolant in crankcase. If no coolant is present, but pressure loss continues:	**11.4**
	If coolant is evident in crankcase, remove cylinder head(s), and check gasket(s). If gaskets are intact, block and cylinder head(s) should be checked for cracks or holes.	
	If the gasket(s) is blown, replace, and purge the crankcase of coolant:	**12.6**
	NOTE: *Occasionally, due to atmospheric and driving conditions, condensation of water can occur in the crankcase. This causes the oil to appear milky white. To remedy, run the engine until hot, and change the oil and oil filter.*	
11.4—Check for combustion leaks into the cooling system: Pressurize the cooling system as above. Start the engine, and observe the pressure gauge. If the needle fluctuates, remove each spark plug wire, one at a time, noting which cylinder(s) reduce or eliminate the fluctuation.	Cylinders which reduce or eliminate the fluctuation, when the spark plug wire is removed, are leaking into the cooling system. Replace the head gasket on the affected cylinder bank(s).	

Pressurizing the cooling system

Test and Procedure	Results and Indications	Proceed to
11.5—Check the radiator pressure cap: Attach a radiator pressure tester to the radiator cap (wet the seal prior to installation). Quickly pump up the pressure, noting the point at which the cap releases.	If the cap releases within ± 1 psi of the specified rating, it is operating properly:	**11.6**
	If the cap releases at more than ± 1 psi of the specified rating, it should be replaced:	**11.6**

Checking radiator pressure cap

Test and Procedure	Results and Indications	Proceed to
11.6—Test the thermostat: Start the engine cold, remove the radiator cap, and insert a thermometer into the radiator. Allow the engine to idle. After a short while, there will be a sudden, rapid increase in coolant temperature. The temperature at which this sharp rise stops is the thermostat opening temperature.	If the thermostat opens at or about the specified temperature:	**11.7**
	If the temperature doesn't increase: (If the temperature increases slowly and gradually, replace the thermostat.)	**11.7**
11.7—Check the water pump: Remove the thermostat elbow and the thermostat, disconnect the coil high tension lead (to prevent starting), and crank the engine momentarily.	If coolant flows, replace the thermostat and retest per 11.6:	**11.6**
	If coolant doesn't flow, reverse flush the cooling system to alleviate any blockage that might exist. If system is not blocked, and coolant will not flow, replace the water pump.	

Section 12—Lubrication
See Chapter 3 for service procedures

Test and Procedure	Results and Indications	Proceed to
12.1—Check the oil pressure gauge or warning light: If the gauge shows low pressure, or the light is on for no obvious reason, remove the oil pressure sender. Install an accurate oil pressure gauge and run the engine momentarily.	If oil pressure builds normally, run engine for a few moments to determine that it is functioning normally, and replace the sender.	—
	If the pressure remains low:	**12.2**
	If the pressure surges:	**12.3**
	If the oil pressure is zero:	**12.3**
12.2—Visually inspect the oil: If the oil is watery or very thin, milky, or foamy, replace the oil and oil filter.	If the oil is normal:	**12.3**
	If after replacing oil the pressure remains low:	**12.3**
	If after replacing oil the pressure becomes normal:	—

Test and Procedure	Results and Indications	Proceed to
12.3—Inspect the oil pressure relief valve and spring, to ensure that it is not sticking or stuck. Remove and thoroughly clean the valve, spring, and the valve body.	If the oil pressure improves: If no improvement is noted:	— **12.4**
12.4—Check to ensure that the oil pump is not cavitating (sucking air instead of oil): See that the crankcase is neither over nor underfull, and that the pickup in the sump is in the proper position and free from sludge.	Fill or drain the crankcase to the proper capacity, and clean the pickup screen in solvent if necessary. If no improvement is noted:	**12.5**
12.5—Inspect the oil pump drive and the oil pump:	If the pump drive or the oil pump appear to be defective, service as necessary and retest per 12.1: If the pump drive and pump appear to be operating normally, the engine should be disassembled to determine where blockage exists:	**12.1** **See Chapter 3**
12.6—Purge the engine of ethylene glycol coolant: Completely drain the crankcase and the oil filter. Obtain a commercial butyl cellosolve base solvent, designated for this purpose, and follow the instructions precisely. Following this, install a new oil filter and refill the crankcase with the proper weight oil. The next oil and filter change should follow shortly thereafter (1000 miles).		

TROUBLESHOOTING EMISSION CONTROL SYSTEMS

See Chapter 4 for procedures applicable to individual emission control systems used on specific combinations of engine/transmission/model.

TROUBLESHOOTING THE CARBURETOR

See Chapter 4 for service procedures

Carburetor problems cannot be effectively isolated unless all other engine systems (particularly ignition and emission) are functioning properly and the engine is properly tuned.

Condition	Possible Cause
Engine cranks, but does not start	1. Improper starting procedure 2. No fuel in tank 3. Clogged fuel line or filter 4. Defective fuel pump 5. Choke valve not closing properly 6. Engine flooded 7. Choke valve not unloading 8. Throttle linkage not making full travel 9. Stuck needle or float 10. Leaking float needle or seat 11. Improper float adjustment
Engine stalls	1. Improperly adjusted idle speed or mixture **Engine hot** 2. Improperly adjusted dashpot 3. Defective or improperly adjusted solenoid 4. Incorrect fuel level in fuel bowl 5. Fuel pump pressure too high 6. Leaking float needle seat 7. Secondary throttle valve stuck open 8. Air or fuel leaks 9. Idle air bleeds plugged or missing 10. Idle passages plugged **Engine Cold** 11. Incorrectly adjusted choke 12. Improperly adjusted fast idle speed 13. Air leaks 14. Plugged idle or idle air passages 15. Stuck choke valve or binding linkage 16. Stuck secondary throttle valves 17. Engine flooding—high fuel level 18. Leaking or misaligned float
Engine hesitates on acceleration	1. Clogged fuel filter 2. Leaking fuel pump diaphragm 3. Low fuel pump pressure 4. Secondary throttle valves stuck, bent or misadjusted 5. Sticking or binding air valve 6. Defective accelerator pump 7. Vacuum leaks 8. Clogged air filter 9. Incorrect choke adjustment (engine cold)
Engine feels sluggish or flat on acceleration	1. Improperly adjusted idle speed or mixture 2. Clogged fuel filter 3. Defective accelerator pump 4. Dirty, plugged or incorrect main metering jets 5. Bent or sticking main metering rods 6. Sticking throttle valves 7. Stuck heat riser 8. Binding or stuck air valve 9. Dirty, plugged or incorrect secondary jets 10. Bent or sticking secondary metering rods. 11. Throttle body or manifold heat passages plugged 12. Improperly adjusted choke or choke vacuum break.
Carburetor floods	1. Defective fuel pump. Pressure too high. 2. Stuck choke valve 3. Dirty, worn or damaged float or needle valve/seat 4. Incorrect float/fuel level 5. Leaking float bowl

Condition	Possible Cause
Engine idles roughly and stalls	1. Incorrect idle speed 2. Clogged fuel filter 3. Dirt in fuel system or carburetor 4. Loose carburetor screws or attaching bolts 5. Broken carburetor gaskets 6. Air leaks 7. Dirty carburetor 8. Worn idle mixture needles 9. Throttle valves stuck open 10. Incorrectly adjusted float or fuel level 11. Clogged air filter
Engine runs unevenly or surges	1. Defective fuel pump 2. Dirty or clogged fuel filter 3. Plugged, loose or incorrect main metering jets or rods 4. Air leaks 5. Bent or sticking main metering rods 6. Stuck power piston 7. Incorrect float adjustment 8. Incorrect idle speed or mixture 9. Dirty or plugged idle system passages 10. Hard, brittle or broken gaskets 11. Loose attaching or mounting screws 12. Stuck or misaligned secondary throttle valves
Poor fuel economy	1. Poor driving habits 2. Stuck choke valve 3. Binding choke linkage 4. Stuck heat riser 5. Incorrect idle mixture 6. Defective accelerator pump 7. Air leaks 8. Plugged, loose or incorrect main metering jets 9. Improperly adjusted float or fuel level 10. Bent, misaligned or fuel-clogged float 11. Leaking float needle seat 12. Fuel leak 13. Accelerator pump discharge ball not seating properly 14. Incorrect main jets
Engine lacks high speed performance or power	1. Incorrect throttle linkage adjustment 2. Stuck or binding power piston 3. Defective accelerator pump 4. Air leaks 5. Incorrect float setting or fuel level 6. Dirty, plugged, worn or incorrect main metering jets or rods 7. Binding or sticking air valve 8. Brittle or cracked gaskets 9. Bent, incorrect or improperly adjusted secondary metering rods 10. Clogged fuel filter 11. Clogged air filter 12. Defective fuel pump

TROUBLESHOOTING FUEL INJECTION PROBLEMS

Each fuel injection system has its own unique components and test procedures, for which it is impossible to generalize. Refer to Chapter 4 of this Repair & Tune-Up Guide for specific test and repair procedures, if the vehicle is equipped with fuel injection.

TROUBLESHOOTING ELECTRICAL PROBLEMS

See Chapter 5 for service procedures

For any electrical system to operate, it must make a complete circuit. This simply means that the power flow from the battery must make a complete circle. When an electrical component is operating, power flows from the battery to the component, passes through the component causing it to perform its function (lighting a light bulb), and then returns to the battery through the ground of the circuit. This ground is usually (but not always) the metal part of the car or truck on which the electrical component is mounted.

Perhaps the easiest way to visualize this is to think of connecting a light bulb with two wires attached to it to the battery. If one of the two wires attached to the light bulb were attached to the negative post of the battery and the other were attached to the positive post of the battery, you would have a complete circuit. Current from the battery would flow to the light bulb, causing it to light, and return to the negative post of the battery.

The normal automotive circuit differs from this simple example in two ways. First, instead of having a return wire from the bulb to the battery, the light bulb returns the current to the battery through the chassis of the vehicle. Since the negative battery cable is attached to the chassis and the chassis is made of electrically conductive metal, the chassis of the vehicle can serve as a ground wire to complete the circuit. Secondly, most automotive circuits contain switches to turn components on and off as required.

Every complete circuit from a power source must include a component which is using the power from the power source. If you were to disconnect the light bulb from the wires and touch the two wires together (don't do this) the power supply wire to the component would be grounded before the normal ground connection for the circuit.

Because grounding a wire from a power source makes a complete circuit—less the required component to use the power—this phenomenon is called a short circuit. Common causes are: broken insulation (exposing the metal wire to a metal part of the car or truck), or a shorted switch.

Some electrical components which require a large amount of current to operate also have a relay in their circuit. Since these circuits carry a large amount of current, the thickness of the wire in the circuit (gauge size) is also greater. If this large wire were connected from the component to the control switch on the instrument panel, and then back to the component, a voltage drop would occur in the circuit. To prevent this potential drop in voltage, an electromagnetic switch (relay) is used. The large wires in the circuit are connected from the battery to one side of the relay, and from the opposite side of the relay to the component. The relay is normally open, preventing current from passing through the circuit. An additional, smaller, wire is connected from the relay to the control switch for the circuit. When the control switch is turned on, it grounds the smaller wire from the relay and completes the circuit. This closes the relay and allows current to flow from the battery to the component. The horn, headlight, and starter circuits are three which use relays.

It is possible for larger surges of current to pass through the electrical system of your car or truck. If this surge of current were to reach an electrical component, it could burn it out. To prevent this, fuses, circuit breakers or fusible links are connected into the current supply wires of most of the major electrical systems. When an electrical current of excessive power passes through the component's fuse, the fuse blows out and breaks the circuit, saving the component from destruction.

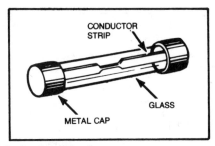

Typical automotive fuse

A circuit breaker is basically a self-repairing fuse. The circuit breaker opens the circuit the same way a fuse does. However, when either the short is removed from the circuit or the surge subsides, the circuit breaker resets itself and does not have to be replaced as a fuse does.

A fuse link is a wire that acts as a fuse. It is normally connected between the starter relay and the main wiring harness. This connection is usually under the hood. The fuse link (if installed) protects all the

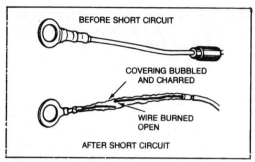

Most fusible links show a charred, melted insulation when they burn out

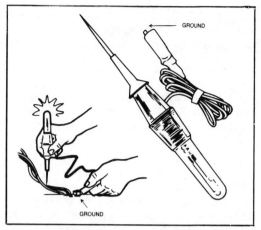

The test light will show the presence of current when touched to a hot wire and grounded at the other end

chassis electrical components, and is the probable cause of trouble when none of the electrical components function, unless the battery is disconnected or dead.

Electrical problems generally fall into one of three areas:

1. The component that is not functioning is not receiving current.

2. The component itself is not functioning.

3. The component is not properly grounded.

The electrical system can be checked with a test light and a jumper wire. A test light is a device that looks like a pointed screwdriver with a wire attached to it and has a light bulb in its handle. A jumper wire is a piece of insulated wire with an alligator clip attached to each end.

If a component is not working, you must follow a systematic plan to determine which of the three causes is the villain.

1. Turn on the switch that controls the inoperable component.

2. Disconnect the power supply wire from the component.

3. Attach the ground wire on the test light to a good metal ground.

4. Touch the probe end of the test light to the end of the power supply wire that was disconnected from the component. If the component is receiving current, the test light will go on.

NOTE: *Some components work only when the ignition switch is turned on.*

If the test light does not go on, then the problem is in the circuit between the battery and the component. This includes all the switches, fuses, and relays in the system. Follow the wire that runs back to the battery. The problem is an open circuit between the

battery and the component. If the fuse is blown and, when replaced, immediately blows again, there is a short circuit in the system which must be located and repaired. If there is a switch in the system, bypass it with a jumper wire. This is done by connecting one end of the jumper wire to the power supply wire into the switch and the other end of the jumper wire to the wire coming out of the switch. If the test light lights with the jumper wire installed, the switch or whatever was bypassed is defective.

NOTE: *Never substitute the jumper wire for the component, since it is required to use the power from the power source.*

5. If the bulb in the test light goes on, then the current is getting to the component that is not working. This eliminates the first of the three possible causes. Connect the power supply wire and connect a jumper wire from the component to a good metal ground. Do this with the switch which controls the component turned on, and also the ignition switch turned on if it is required for the component to work. If the component works with the jumper wire installed, then it has a bad ground. This is usually caused by the metal area on which the component mounts to the chassis being coated with some type of foreign matter.

6. If neither test located the source of the trouble, then the component itself is defective. Remember that for any electrical system to work, all connections must be clean and tight.

Troubleshooting Basic Turn Signal and Flasher Problems
See Chapter 5 for service procedures

Most problems in the turn signals or flasher system can be reduced to defective flashers or bulbs, which are easily replaced. Occasionally, the turn signal switch will prove defective.

F = Front   R = Rear   ● = Lights off   ○ = Lights on

Condition		Possible Cause
Turn signals light, but do not flash	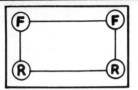	Defective flasher
No turn signals light on either side	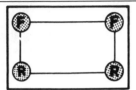	Blown fuse. Replace if defective. Defective flasher. Check by substitution. Open circuit, short circuit or poor ground.
Both turn signals on one side don't work	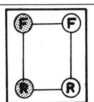	Bad bulbs. Bad ground in both (or either) housings.
One turn signal light on one side doesn't work	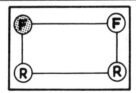	Defective bulb. Corrosion in socket. Clean contacts. Poor ground at socket.
Turn signal flashes too fast or too slowly	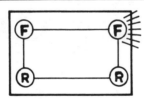	Check any bulb on the side flashing too fast. A heavy-duty bulb is probably installed in place of a regular bulb. Check the bulb flashing too slowly. A standard bulb was probably installed in place of a heavy-duty bulb. Loose connections or corrosion at the bulb socket.
Indicator lights don't work in either direction		Check if the turn signals are working. Check the dash indicator lights. Check the flasher by substitution.
One indicator light doesn't light		On systems with one dash indicator: See if the lights work on the same side. Often the filaments have been reversed in systems combining stoplights with taillights and turn signals. Check the flasher by substitution. On systems with two indicators: Check the bulbs on the same side. Check the indicator light bulb. Check the flasher by substitution.

Troubleshooting Lighting Problems

See Chapter 5 for service procedures

Condition	Possible Cause
One or more lights don't work, but others do	1. Defective bulb(s) 2. Blown fuse(s) 3. Dirty fuse clips or light sockets 4. Poor ground circuit
Lights burn out quickly	1. Incorrect voltage regulator setting or defective regulator 2. Poor battery/alternator connections
Lights go dim	1. Low/discharged battery 2. Alternator not charging 3. Corroded sockets or connections 4. Low voltage output
Lights flicker	1. Loose connection 2. Poor ground. (Run ground wire from light housing to frame) 3. Circuit breaker operating (short circuit)
Lights "flare"—Some flare is normal on acceleration—If excessive, see "Lights Burn Out Quickly"	High voltage setting
Lights glare—approaching drivers are blinded	1. Lights adjusted too high 2. Rear springs or shocks sagging 3. Rear tires soft

Troubleshooting Dash Gauge Problems

Most problems can be traced to a defective sending unit or faulty wiring. Occasionally, the gauge itself is at fault. See Chapter 5 for service procedures.

Condition	Possible Cause

COOLANT TEMPERATURE GAUGE

Gauge reads erratically or not at all	1. Loose or dirty connections 2. Defective sending unit. 3. Defective gauge. To test a bi-metal gauge, remove the wire from the sending unit. Ground the wire for an instant. If the gauge registers, replace the sending unit. To test a magnetic gauge, disconnect the wire at the sending unit. With ignition ON gauge should register COLD. Ground the wire; gauge should register HOT.

AMMETER GAUGE—TURN HEADLIGHTS ON (DO NOT START ENGINE). NOTE REACTION

Ammeter shows charge Ammeter shows discharge Ammeter does not move	1. Connections reversed on gauge 2. Ammeter is OK 3. Loose connections or faulty wiring 4. Defective gauge

Condition	Possible Cause

OIL PRESSURE GAUGE

Condition	Possible Cause
Gauge does not register or is inaccurate	1. On mechanical gauge, Bourdon tube may be bent or kinked. 2. Low oil pressure. Remove sending unit. Idle the engine briefly. If no oil flows from sending unit hole, problem is in engine. 3. Defective gauge. Remove the wire from the sending unit and ground it for an instant with the ignition ON. A good gauge will go to the top of the scale. 4. Defective wiring. Check the wiring to the gauge. If it's OK and the gauge doesn't register when grounded, replace the gauge. 5. Defective sending unit.

ALL GAUGES

Condition	Possible Cause
All gauges do not operate All gauges read low or erratically All gauges pegged	1. Blown fuse 2. Defective instrument regulator 3. Defective or dirty instrument voltage regulator 4. Loss of ground between instrument voltage regulator and frame 5. Defective instrument regulator

WARNING LIGHTS

Condition	Possible Cause
Light(s) do not come on when ignition is ON, but engine is not started Light comes on with engine running	1. Defective bulb 2. Defective wire 3. Defective sending unit. Disconnect the wire from the sending unit and ground it. Replace the sending unit if the light comes on with the ignition ON. 4. Problem in individual system 5. Defective sending unit

Troubleshooting Clutch Problems

It is false economy to replace individual clutch components. The pressure plate, clutch plate and throwout bearing should be replaced as a set, and the flywheel face inspected, whenever the clutch is overhauled. See Chapter 6 for service procedures.

Condition	Possible Cause
Clutch chatter	1. Grease on driven plate (disc) facing 2. Binding clutch linkage or cable 3. Loose, damaged facings on driven plate (disc) 4. Engine mounts loose 5. Incorrect height adjustment of pressure plate release levers 6. Clutch housing or housing to transmission adapter misalignment 7. Loose driven plate hub
Clutch grabbing	1. Oil, grease on driven plate (disc) facing 2. Broken pressure plate 3. Warped or binding driven plate. Driven plate binding on clutch shaft
Clutch slips	1. Lack of lubrication in clutch linkage or cable (linkage or cable binds, causes incomplete engagement) 2. Incorrect pedal, or linkage adjustment 3. Broken pressure plate springs 4. Weak pressure plate springs 5. Grease on driven plate facings (disc)

Troubleshooting Clutch Problems (cont.)

Condition	Possible Cause
Incomplete clutch release	1. Incorrect pedal or linkage adjustment or linkage or cable binding 2. Incorrect height adjustment on pressure plate release levers 3. Loose, broken facings on driven plate (disc) 4. Bent, dished, warped driven plate caused by overheating
Grinding, whirring grating noise when pedal is depressed	1. Worn or defective throwout bearing 2. Starter drive teeth contacting flywheel ring gear teeth. Look for milled or polished teeth on ring gear.
Squeal, howl, trumpeting noise when pedal is being released (occurs during first inch to inch and one-half of pedal travel)	Pilot bushing worn or lack of lubricant. If bushing appears OK, polish bushing with emery cloth, soak lube wick in oil, lube bushing with oil, apply film of chassis grease to clutch shaft pilot hub, reassemble. NOTE: Bushing wear may be due to misalignment of clutch housing or housing to transmission adapter
Vibration or clutch pedal pulsation with clutch disengaged (pedal fully depressed)	1. Worn or defective engine transmission mounts 2. Flywheel run out. (Flywheel run out at face not to exceed 0.005") 3. Damaged or defective clutch components

Troubleshooting Manual Transmission Problems
See Chapter 6 for service procedures

Condition	Possible Cause
Transmission jumps out of gear	1. Misalignment of transmission case or clutch housing. 2. Worn pilot bearing in crankshaft. 3. Bent transmission shaft. 4. Worn high speed sliding gear. 5. Worn teeth or end-play in clutch shaft. 6. Insufficient spring tension on shifter rail plunger. 7. Bent or loose shifter fork. 8. Gears not engaging completely. 9. Loose or worn bearings on clutch shaft or mainshaft. 10. Worn gear teeth. 11. Worn or damaged detent balls.
Transmission sticks in gear	1. Clutch not releasing fully. 2. Burred or battered teeth on clutch shaft, or sliding sleeve. 3. Burred or battered transmission mainshaft. 4. Frozen synchronizing clutch. 5. Stuck shifter rail plunger. 6. Gearshift lever twisting and binding shifter rail. 7. Battered teeth on high speed sliding gear or on sleeve. 8. Improper lubrication, or lack of lubrication. 9. Corroded transmission parts. 10. Defective mainshaft pilot bearing. 11. Locked gear bearings will give same effect as stuck in gear.
Transmission gears will not synchronize	1. Binding pilot bearing on mainshaft, will synchronize in high gear only. 2. Clutch not releasing fully. 3. Detent spring weak or broken. 4. Weak or broken springs under balls in sliding gear sleeve. 5. Binding bearing on clutch shaft, or binding countershaft. 6. Binding pilot bearing in crankshaft. 7. Badly worn gear teeth. 8. Improper lubrication. 9. Constant mesh gear not turning freely on transmission mainshaft. Will synchronize in that gear only.

Condition	Possible Cause
Gears spinning when shifting into gear from neutral	1. Clutch not releasing fully. 2. In some cases an extremely light lubricant in transmission will cause gears to continue to spin for a short time after clutch is released. 3. Binding pilot bearing in crankshaft.
Transmission noisy in all gears	1. Insufficient lubricant, or improper lubricant. 2. Worn countergear bearings. 3. Worn or damaged main drive gear or countergear. 4. Damaged main drive gear or mainshaft bearings. 5. Worn or damaged countergear anti-lash plate.
Transmission noisy in neutral only	1. Damaged main drive gear bearing. 2. Damaged or loose mainshaft pilot bearing. 3. Worn or damaged countergear anti-lash plate. 4. Worn countergear bearings.
Transmission noisy in one gear only	1. Damaged or worn constant mesh gears. 2. Worn or damaged countergear bearings. 3. Damaged or worn synchronizer.
Transmission noisy in reverse only	1. Worn or damaged reverse idler gear or idler bushing. 2. Worn or damaged mainshaft reverse gear. 3. Worn or damaged reverse countergear. 4. Damaged shift mechanism.

TROUBLESHOOTING AUTOMATIC TRANSMISSION PROBLEMS

Keeping alert to changes in the operating characteristics of the transmission (changing shift points, noises, etc.) can prevent small problems from becoming large ones. If the problem cannot be traced to loose bolts, fluid level, misadjusted linkage, clogged filters or similar problems, you should probably seek professional service.

Transmission Fluid Indications

The appearance and odor of the transmission fluid can give valuable clues to the overall condition of the transmission. Always note the appearance of the fluid when you check the fluid level or change the fluid. Rub a small amount of fluid between your fingers to feel for grit and smell the fluid on the dipstick.

If the fluid appears:	It indicates:
Clear and red colored	Normal operation
Discolored (extremely dark red or brownish) or smells burned	Band or clutch pack failure, usually caused by an overheated transmission. Hauling very heavy loads with insufficient power or failure to change the fluid often result in overheating. Do not confuse this appearance with newer fluids that have a darker red color and a strong odor (though not a burned odor).
Foamy or aerated (light in color and full of bubbles)	1. The level is too high (gear train is churning oil) 2. An internal air leak (air is mixing with the fluid). Have the transmission checked professionally.
Solid residue in the fluid	Defective bands, clutch pack or bearings. Bits of band material or metal abrasives are clinging to the dipstick. Have the transmission checked professionally.
Varnish coating on the dipstick	The transmission fluid is overheating

TROUBLESHOOTING DRIVE AXLE PROBLEMS

First, determine when the noise is most noticeable.

Drive Noise: Produced under vehicle acceleration.

Coast Noise: Produced while coasting with a closed throttle.

Float Noise: Occurs while maintaining constant speed (just enough to keep speed constant) on a level road.

External Noise Elimination

It is advisable to make a thorough road test to determine whether the noise originates in the rear axle or whether it originates from the tires, engine, transmission, wheel bearings or road surface. Noise originating from other places cannot be corrected by servicing the rear axle.

ROAD NOISE

Brick or rough surfaced concrete roads produce noises that seem to come from the rear axle. Road noise is usually identical in Drive or Coast and driving on a different type of road will tell whether the road is the problem.

TIRE NOISE

Tire noise can be mistaken as rear axle noise, even though the tires on the front are at fault. Snow tread and mud tread tires or tires worn unevenly will frequently cause vibrations which seem to originate elsewhere; *temporarily, and for test purposes only,* inflate the tires to 40–50 lbs. This will significantly alter the noise produced by the tires, but will not alter noise from the rear axle. Noises from the rear axle will normally cease at speeds below 30 mph on coast, while tire noise will continue at lower tone as speed is decreased. The rear axle noise will usually change from drive conditions to coast conditions, while tire noise will not. Do not forget to lower the tire pressure to normal after the test is complete.

ENGINE/TRANSMISSION NOISE

Determine at what speed the noise is most pronounced, then stop in a quiet place. With the transmission in Neutral, run the engine through speeds corresponding to road speeds where the noise was noticed. Noises produced with the vehicle standing still are coming from the engine or transmission.

FRONT WHEEL BEARINGS

Front wheel bearing noises, sometimes confused with rear axle noises, will not change when comparing drive and coast conditions. While holding the speed steady, lightly apply the footbrake. This will often cause wheel bearing noise to lessen, as some of the weight is taken off the bearing. Front wheel bearings are easily checked by jacking up the wheels and spinning the wheels. Shaking the wheels will also determine if the wheel bearings are excessively loose.

REAR AXLE NOISES

Eliminating other possible sources can narrow the cause to the rear axle, which normally produces noise from worn gears or bearings. Gear noises tend to peak in a narrow speed range, while bearing noises will usually vary in pitch with engine speeds.

Noise Diagnosis

The Noise Is:	Most Probably Produced By:
1. Identical under Drive or Coast	Road surface, tires or front wheel bearings
2. Different depending on road surface	Road surface or tires
3. Lower as speed is lowered	Tires
4. Similar when standing or moving	Engine or transmission
5. A vibration	Unbalanced tires, rear wheel bearing, unbalanced driveshaft or worn U-joint
6. A knock or click about every two tire revolutions	Rear wheel bearing
7. Most pronounced on turns	Damaged differential gears
8. A steady low-pitched whirring or scraping, starting at low speeds	Damaged or worn pinion bearing
9. A chattering vibration on turns	Wrong differential lubricant or worn clutch plates (limited slip rear axle)
10. Noticed only in Drive, Coast or Float conditions	Worn ring gear and/or pinion gear

Troubleshooting Steering & Suspension Problems

Condition	Possible Cause
Hard steering (wheel is hard to turn)	1. Improper tire pressure 2. Loose or glazed pump drive belt 3. Low or incorrect fluid 4. Loose, bent or poorly lubricated front end parts 5. Improper front end alignment (excessive caster) 6. Bind in steering column or linkage 7. Kinked hydraulic hose 8. Air in hydraulic system 9. Low pump output or leaks in system 10. Obstruction in lines 11. Pump valves sticking or out of adjustment 12. Incorrect wheel alignment
Loose steering (too much play in steering wheel)	1. Loose wheel bearings 2. Faulty shocks 3. Worn linkage or suspension components 4. Loose steering gear mounting or linkage points 5. Steering mechanism worn or improperly adjusted 6. Valve spool improperly adjusted 7. Worn ball joints, tie-rod ends, etc.
Veers or wanders (pulls to one side with hands off steering wheel)	1. Improper tire pressure 2. Improper front end alignment 3. Dragging or improperly adjusted brakes 4. Bent frame 5. Improper rear end alignment 6. Faulty shocks or springs 7. Loose or bent front end components 8. Play in Pitman arm 9. Steering gear mountings loose 10. Loose wheel bearings 11. Binding Pitman arm 12. Spool valve sticking or improperly adjusted 13. Worn ball joints
Wheel oscillation or vibration transmitted through steering wheel	1. Low or uneven tire pressure 2. Loose wheel bearings 3. Improper front end alignment 4. Bent spindle 5. Worn, bent or broken front end components 6. Tires out of round or out of balance 7. Excessive lateral runout in disc brake rotor 8. Loose or bent shock absorber or strut
Noises (see also "Troubleshooting Drive Axle Problems")	1. Loose belts 2. Low fluid, air in system 3. Foreign matter in system 4. Improper lubrication 5. Interference or chafing in linkage 6. Steering gear mountings loose 7. Incorrect adjustment or wear in gear box 8. Faulty valves or wear in pump 9. Kinked hydraulic lines 10. Worn wheel bearings
Poor return of steering	1. Over-inflated tires 2. Improperly aligned front end (excessive caster) 3. Binding in steering column 4. No lubrication in front end 5. Steering gear adjusted too tight
Uneven tire wear (see "How To Read Tire Wear")	1. Incorrect tire pressure 2. Improperly aligned front end 3. Tires out-of-balance 4. Bent or worn suspension parts

HOW TO READ TIRE WEAR

The way your tires wear is a good indicator of other parts of the suspension. Abnormal wear patterns are often caused by the need for simple tire maintenance, or for front end alignment.

Excessive wear at the center of the tread indicates that the air pressure in the tire is consistently too high. The tire is riding on the center of the tread and wearing it prematurely. Occasionally, this wear pattern can result from outrageously wide tires on narrow rims. The cure for this is to replace either the tires or the wheels.

This type of wear usually results from consistent under-inflation. When a tire is under-inflated, there is too much contact with the road by the outer treads, which wear prematurely. When this type of wear occurs, and the tire pressure is known to be consistently correct, a bent or worn steering component or the need for wheel alignment could be indicated.

Feathering is a condition when the edge of each tread rib develops a slightly rounded edge on one side and a sharp edge on the other. By running your hand over the tire, you can usually feel the sharper edges before you'll be able to see them. The most common causes of feathering are incorrect toe-in setting or deteriorated bushings in the front suspension.

When an inner or outer rib wears faster than the rest of the tire, the need for wheel alignment is indicated. There is excessive camber in the front suspension, causing the wheel to lean too much putting excessive load on one side of the tire. Misalignment could also be due to sagging springs, worn ball joints, or worn control arm bushings. Be sure the vehicle is loaded the way it's normally driven when you have the wheels aligned.

Cups or scalloped dips appearing around the edge of the tread almost always indicate worn (sometimes bent) suspension parts. Adjustment of wheel alignment alone will seldom cure the problem. Any worn component that connects the wheel to the suspension can cause this type of wear. Occasionally, wheels that are out of balance will wear like this, but wheel imbalance usually shows up as bald spots between the outside edges and center of the tread.

Second-rib wear is usually found only in radial tires, and appears where the steel belts end in relation to the tread. It can be kept to a minimum by paying careful attention to tire pressure and frequently rotating the tires. This is often considered normal wear but excessive amounts indicate that the tires are too wide for the wheels.

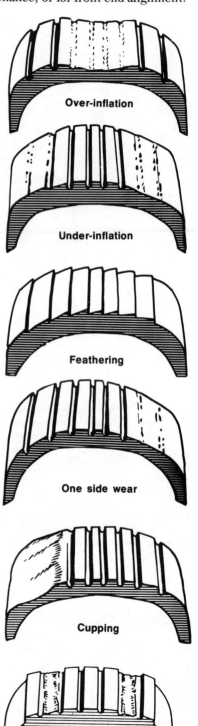

Over-inflation

Under-inflation

Feathering

One side wear

Cupping

Second-rib wear

Troubleshooting Disc Brake Problems

Condition	Possible Cause
Noise—groan—brake noise emanating when slowly releasing brakes (creep-groan)	Not detrimental to function of disc brakes—no corrective action required. (This noise may be eliminated by slightly increasing or decreasing brake pedal efforts.)
Rattle—brake noise or rattle emanating at low speeds on rough roads, (front wheels only).	1. Shoe anti-rattle spring missing or not properly positioned. 2. Excessive clearance between shoe and caliper. 3. Soft or broken caliper seals. 4. Deformed or misaligned disc. 5. Loose caliper.
Scraping	1. Mounting bolts too long. 2. Loose wheel bearings. 3. Bent, loose, or misaligned splash shield.
Front brakes heat up during driving and fail to release	1. Operator riding brake pedal. 2. Stop light switch improperly adjusted. 3. Sticking pedal linkage. 4. Frozen or seized piston. 5. Residual pressure valve in master cylinder. 6. Power brake malfunction. 7. Proportioning valve malfunction.
Leaky brake caliper	1. Damaged or worn caliper piston seal. 2. Scores or corrosion on surface of cylinder bore.
Grabbing or uneven brake action— Brakes pull to one side	1. Causes listed under "Brakes Pull". 2. Power brake malfunction. 3. Low fluid level in master cylinder. 4. Air in hydraulic system. 5. Brake fluid, oil or grease on linings. 6. Unmatched linings. 7. Distorted brake pads. 8. Frozen or seized pistons. 9. Incorrect tire pressure. 10. Front end out of alignment. 11. Broken rear spring. 12. Brake caliper pistons sticking. 13. Restricted hose or line. 14. Caliper not in proper alignment to braking disc. 15. Stuck or malfunctioning metering valve. 16. Soft or broken caliper seals. 17. Loose caliper.
Brake pedal can be depressed without braking effect	1. Air in hydraulic system or improper bleeding procedure. 2. Leak past primary cup in master cylinder. 3. Leak in system. 4. Rear brakes out of adjustment. 5. Bleeder screw open.
Excessive pedal travel	1. Air, leak, or insufficient fluid in system or caliper. 2. Warped or excessively tapered shoe and lining assembly. 3. Excessive disc runout. 4. Rear brake adjustment required. 5. Loose wheel bearing adjustment. 6. Damaged caliper piston seal. 7. Improper brake fluid (boil). 8. Power brake malfunction. 9. Weak or soft hoses.

Troubleshooting Disc Brake Problems (cont.)

Condition	Possible Cause
Brake roughness or chatter (pedal pumping)	1. Excessive thickness variation of braking disc. 2. Excessive lateral runout of braking disc. 3. Rear brake drums out-of-round. 4. Excessive front bearing clearance.
Excessive pedal effort	1. Brake fluid, oil or grease on linings. 2. Incorrect lining. 3. Frozen or seized pistons. 4. Power brake malfunction. 5. Kinked or collapsed hose or line. 6. Stuck metering valve. 7. Scored caliper or master cylinder bore. 8. Seized caliper pistons.
Brake pedal fades (pedal travel increases with foot on brake)	1. Rough master cylinder or caliper bore. 2. Loose or broken hydraulic lines/connections. 3. Air in hydraulic system. 4. Fluid level low. 5. Weak or soft hoses. 6. Inferior quality brake shoes or fluid. 7. Worn master cylinder piston cups or seals.

Troubleshooting Drum Brakes

Condition	Possible Cause
Pedal goes to floor	1. Fluid low in reservoir. 2. Air in hydraulic system. 3. Improperly adjusted brake. 4. Leaking wheel cylinders. 5. Loose or broken brake lines. 6. Leaking or worn master cylinder. 7. Excessively worn brake lining.
Spongy brake pedal	1. Air in hydraulic system. 2. Improper brake fluid (low boiling point). 3. Excessively worn or cracked brake drums. 4. Broken pedal pivot bushing.
Brakes pulling	1. Contaminated lining. 2. Front end out of alignment. 3. Incorrect brake adjustment. 4. Unmatched brake lining. 5. Brake drums out of round. 6. Brake shoes distorted. 7. Restricted brake hose or line. 8. Broken rear spring. 9. Worn brake linings. 10. Uneven lining wear. 11. Glazed brake lining. 12. Excessive brake lining dust. 13. Heat spotted brake drums. 14. Weak brake return springs. 15. Faulty automatic adjusters. 16. Low or incorrect tire pressure.

Condition	Possible Cause
Squealing brakes	1. Glazed brake lining. 2. Saturated brake lining. 3. Weak or broken brake shoe retaining spring. 4. Broken or weak brake shoe return spring. 5. Incorrect brake lining. 6. Distorted brake shoes. 7. Bent support plate. 8. Dust in brakes or scored brake drums. 9. Linings worn below limit. 10. Uneven brake lining wear. 11. Heat spotted brake drums.
Chirping brakes	1. Out of round drum or eccentric axle flange pilot.
Dragging brakes	1. Incorrect wheel or parking brake adjustment. 2. Parking brakes engaged or improperly adjusted. 3. Weak or broken brake shoe return spring. 4. Brake pedal binding. 5. Master cylinder cup sticking. 6. Obstructed master cylinder relief port. 7. Saturated brake lining. 8. Bent or out of round brake drum. 9. Contaminated or improper brake fluid. 10. Sticking wheel cylinder pistons. 11. Driver riding brake pedal. 12. Defective proportioning valve. 13. Insufficient brake shoe lubricant.
Hard pedal	1. Brake booster inoperative. 2. Incorrect brake lining. 3. Restricted brake line or hose. 4. Frozen brake pedal linkage. 5. Stuck wheel cylinder. 6. Binding pedal linkage. 7. Faulty proportioning valve.
Wheel locks	1. Contaminated brake lining. 2. Loose or torn brake lining. 3. Wheel cylinder cups sticking. 4. Incorrect wheel bearing adjustment. 5. Faulty proportioning valve.
Brakes fade (high speed)	1. Incorrect lining. 2. Overheated brake drums. 3. Incorrect brake fluid (low boiling temperature). 4. Saturated brake lining. 5. Leak in hydraulic system. 6. Faulty automatic adjusters.
Pedal pulsates	1. Bent or out of round brake drum.
Brake chatter and shoe knock	1. Out of round brake drum. 2. Loose support plate. 3. Bent support plate. 4. Distorted brake shoes. 5. Machine grooves in contact face of brake drum (Shoe Knock). 6. Contaminated brake lining. 7. Missing or loose components. 8. Incorrect lining material. 9. Out-of-round brake drums. 10. Heat spotted or scored brake drums. 11. Out-of-balance wheels.

Troubleshooting Drum Brakes (cont.)

Condition	Possible Cause
Brakes do not self adjust	1. Adjuster screw frozen in thread. 2. Adjuster screw corroded at thrust washer. 3. Adjuster lever does not engage star wheel. 4. Adjuster installed on wrong wheel.
Brake light glows	1. Leak in the hydraulic system. 2. Air in the system. 3. Improperly adjusted master cylinder pushrod. 4. Uneven lining wear. 5. Failure to center combination valve or proportioning valve.

Index